Wicked Philosophy

Wicked Philosophy

Philosophy of Science and Vision Development for Complex Problems

Coyan Tromp

Amsterdam University Press

Volume 5 of the Series Perspectives on Interdisciplinarity

Cover design and lay-out: Matterhorn Amsterdam

ISBN   978 94 6298 877 4
e-ISBN   978 90 4854 109 6 (pdf)
DOI   10.5117/9789462988774
NUR   734

Contents

Preface

A decade ago, I was recruited by the University of Amsterdam to help design and develop Future Planet Studies, an interdisciplinary Bachelor's programme that takes as its point of departure the challenges related to safeguarding a sustainable future for humankind on our planet. The issues involved – such as climate change, energy, food and water demand – are so complex and so difficult to resolve that they have been called persistent or 'wicked' problems. Assembling experts from various fields of science for the programme was a challenging yet rewarding task. We needed to gather together human geographers and political scientists as well as earth scientists and ecologists, not to mention experts on the economic and communications aspects of the various issues. My background in philosophy of science and psychology proved useful – it helped me understand the different positions held by the professors and lecturers. Realising that their positions could vary significantly, I paid close attention to the different ways they perceived the challenges, both with regard to the actual problems and to the development of a completely new curriculum that broke with conventional academic rules and regulations. Only by taking these various perspectives into account could we hope to develop a suitable curriculum, i.e. an educational programme designed to address the complex problems that represent the central focus of this programme.

When I was given responsibility for a course on philosophy of science and vision development within the programme, I faced similar challenges. It was difficult to find appropriate material. Traditional books on philosophy of science offer relatively few answers to many of today's questions. Either they focus on the natural sciences or the social sciences or the humanities; they rarely cover more than one domain. Yet we need to examine all of these and how they interact with each other to be able to find the necessary integrated approaches to contemporary challenges. Moreover, the available books usually focus on the history of the philosophy of science, while our concern is to tackle the urgent complex problems of today and tomorrow. Obviously, knowledge and experience gained in the past provide a crucial foundation for our efforts today, yet our explicit focus is the future. We need the latest insights and innovative visions to inspire us to find viable solutions. I undertook to write this book in the hope that I could help to fill that gap and meet these needs.

The big question is: how do we deal with the complex issues that confront us? Traditionally, science has aimed to provide answers to questions relating to the world in which we live. Yet can science help provide answers to the highly complex issues that we are currently faced with? Can it offer insights and explanations and help find solutions to our urgent practical needs? Specifically: what is the role and the value of natural sciences, and what is the role of the social sciences and the humanities? How can we combine and integrate knowledge gained in diverse disciplines to provide insights, explanations and solutions? And what role does philosophy of science play in this process? These are the questions at the core of this book: a discussion of how philosophy, especially philosophy of science, can help us learn to deal with complex issues. In the end, 'wicked' problems need an accompanying 'wicked' philosophy – hence the title.

We start in chapter 1 with the good news that in response to the persistent complex problems, a new, concurrent way of thinking has evolved: complexity thinking. We investigate what makes this new approach different and where it diverges from conventional approaches in science. We also examine whether it really is geared to the particular character of complex problems. We discover that every approach inevitably rests on certain presuppositions: basic assumptions that determine how the research process is designed yet possess no scientific legitimacy in themselves. As we analyse these assumptions and reflect on their implications, the contours of a 'wicked' philosophy of science take shape, showing the conditions that science must fulfil to meet today's demands.

In chapter 2, we take a closer look at both traditional, existing approaches in science and the new approach now evolving. We investigate the various functions these new approaches can perform in terms of understanding, finding explanations, providing solutions and realising social change. We also address major criticisms that have been raised regarding each scientific approach. Our analysis shows that we need to combine all the available approaches to gain the necessary range and depth of insight into the complex issues of our time.

This task is pursued in chapter 3, where we try to build bridges between the various approaches developed in the natural sciences on the one hand and in the social sciences and the humanities on the other. We propose complexity thinking as a meta-position in which the best features of the underlying approaches are integrated and the disadvantages of each approach are avoided. This is possible when systems thinking and research into physical and social structures are accompanied by inquiry into agency. It also requires that the interrelation between structure and action are explicitly taken into account. We investigate how design thinking can help shape this aspect regarding action and how it can enhance the implementation of policy strategies aimed at solving urgent problems.

This is all the more relevant in chapter 4, where we discuss whether science always produces the most rational solutions. While it is defined as a rational learning process, science has occasionally produced knowledge and technologies that have generated less-than-optimal institutional arrangements and systems. Some if not most of the 'wicked' problems we face are essentially unintended, unwanted side effects of well-meant scientific solutions. So we must conclude that rational decisions do not by definition coincide with wise decisions. We examine what causes this discrepancy and what we can do to reduce the risk of creating precarious situations in the future.

In chapter 5, we review the implications of a 'wicked' philosophy of science for methodology and society, and we explore how to develop sufficiently robust knowledge to enable us to find solutions to persistent, complex problems. We start with a brief recap of the various functions of the different types of research in complex problem-solving. We pay special attention to showing how research projects can be designed to enhance the engagement of scientific researchers and other stakeholders in real-life complexity. Then we look at whether science generates progress in society, and if so, under what conditions. Arguing that traditional standards of scientific knowledge are outdated, we discuss quality criteria that may be better suited to science today. Finally, we evaluate what all this means for the institutional structure of society and for researchers who need to be able to deal with complexity.

In chapter 6, we conclude by exploring the role of vision in the search for robust solutions to urgent problems. Science-based vision can help steer society in the direction of a sustainable future: vision is an indispensable source of inspiration, and it encourages us to take action and points us in the right direction. But what does vision involve? How does vision differ from scientific insight? We end with an assessment of the current situation in scientific education. How are we doing, viewed from the perspective of the challenges we face and the demands these place on today's students and future scientists? Are we on the right track? Or are we on the brink of a crisis that will demand a fundamental change in the way we organise science and society to meet the challenge of the 'wicked' problems? We present the views of some of the world's most prominent scientists and leaders and trust that by this point in the book your insight and critical skills have been sufficiently developed to enable you to determine your own position in this contemporary debate.

We help you to do this by introducing a reflexive way of thinking about complex problems that incorporates both the structural systems perspective and the actor's perspective. We also provide key conceptual tools to enable you to engage in projects involving complex problems, helping you monitor intended and possibly unintended effects of interventions designed to produce solutions to today's challenges. These tools consist of a broad range of theories and concepts to deal with the complexity of 'wicked' problems, i.e., to gain more understanding and provide ideas about how to tackle them. To help you handle the tools, definitions of the cursified key concepts are provided in a glossary at the end of the book. This glossary also serves as an index.

Since it is impossible to study the whole world and its surroundings simultaneously, the book focuses on challenges at the interface of humankind and planet Earth.

One challenge in particular – the food issue – will serve as our central case study. Various aspects of the food issue are used as concrete illustrations of the general philosophical issues that are addressed. In 2050, we will need to feed an expected global population of 9.7 billion in sustainable ways, i.e. ways that the planet has the capacity to maintain. Here we examine the type of issues this involves and the solutions proposed. These examples are presented in a series of boxes. They can help us discover how to deal not only with the food issue but also with relevant related problems. This way, the chosen example can help us deal with comparable complex problems, even with 'wicked' problems outside the chosen intersection of humankind and Earth.

I would like to express my gratitude to the Institute of Interdisciplinary Studies at the University of Amsterdam for giving me the opportunity to write this book. My thanks are also due to Njal van Woerden, Karel van Dam and Lucas van der Zee for their constructive feedback on the early drafts. I am grateful to Huub Dijstelbloem for the seed he planted in my head when he advised me to emphasise the new, more visionary developments, which encouraged me to restructure the book fundamentally a year later. My gratitude also goes to our former scientific director Steph Menken, who provided valuable comments and suggestions on some of the biological examples cited here, to my colleague John van Boxel for doing the same for the Milankovitch' example, and our current scientific director Henk-Jan Honing, who offered useful suggestions in a later phase of the writing process. My philosophical buddy Machiel Keestra, with whom I had the opportunity to exchange ideas while writing this work, provided stimulating boosts when I needed it. I am also grateful to John Grin for having shared his inspiring thoughts, particularly with regard to the action and design dimensions, which have hopefully received their proper due in the book. Jeroen van Dongen's '(no) nonsense!' comments were also very valuable – they forced me to take a critical look at my text and helped me re-examine and present the most valuable points more clearly. And Chunglin Kwa's friendly yet rigorous review helped me 'kill my darlings' and put the finishing touch on the penultimate draft.

My reviewers have helped me avoid flawed interpretations and have corrected errors easily made in the terrain covered by this book. If any mistakes remain, these are due to my own limited capacity to oversee everything that has been said and written about the broad range of topics discussed here. I hope that you, the reader, can appreciate the effort I have made to offer an interdisciplinary and to some extent even transdisciplinary philosophy of science, and that this step-by-step introduction to the issues of 'wicked' philosophy will enhance your ability to deal with the pressing, complex, 'wicked' problems of our times.

Coyan Tromp
Amsterdam, June 2018
j.c.tromp@uva.nl

1 Twenty-First-Century Science

The general aim of philosophy, in one of the most simple and straightforward definitions, is to try to find out how we can come to grips with reality. Philosophers are interested in finding out how we can develop knowledge about the world we live in and how we can describe and explain the phenomena in that world in order to gain deeper insights. They are also interested in developing solutions to tackle problematic issues and in reflecting on the social and ethical consequences. Yet in a world where we are faced with phenomena characterised by immense complexity, formulating an answer to that question is neither simple nor straightforward. Indeed, the complex problems of today pose a huge challenge to scientists. Trying to understand them and formulating theories to explain how they work is no small matter. Providing ideas for possible solutions and anticipating their effects is even more difficult. To better understand what the challenge actually entails, we start this first chapter by taking a closer look at the nature of complex problems. These turn out to have a rather peculiar character, different in important respects from problems encountered in earlier stages of human history.

Next, we examine how science can help satisfy our current knowledge needs. We explore how modern science has evolved and the vision of knowledge production it entails. We review the useful insights and explanations it has produced. Yet as we investigate its foundations, it becomes clear that the *assumptions* underlying the standard scientific approach cannot simply be taken for granted. For one, not everyone agrees that they apply to every area of science. Another reason is that these assumptions fail when employed in the context of complex issues that scientists face today.

This gives cause for explicitly reflecting on the role of assumptions in science in general. In doing so, we find that they are unavoidable. In normal, everyday science, assumptions exercise an enormous influence on the choices researchers make in their search for scientific explanations. By showing how scientists are guided in their work by the overarching mind frame of their time – the zeitgeist – we hope to clarify the vital role that philosophy of science plays in explaining the ways of the scientific enterprise.

We end the chapter with an assessment of the state of science today. Does the advent of complexity thinking offer us a new way to tackle today's urgent issues? Does it present a way of thinking that allows us to capture and deal with the specific, peculiar characteristics of complex problems? Let's find out.

1.1 'Wicked' Problems: The Great Challenges of Our Times

Climate change, energy, food and water demand, sustainability, security, urbanisation, migration and mobility are just some of the issues that have turned out to be both persistent and resistant to easy solutions. They are not merely complicated, meaning that they have many components with many specific functions; they are truly complex: they are multi-level phenomena involving a multiplicity of mutually interacting actors and factors, and their functions cannot be localised in any specific component. We cannot simply combine the pieces to find out how the whole works. Moreover, many of these issues are intrinsically connected to each other on different, higher levels. How can we understand these issues? Will we ever be able to grasp how the underlying phenomena work and interrelate? It is difficult enough to agree on a definition of these problems, let alone reach a consensus about the best solutions. That is why these complex issues and the challenges they involve have been called 'wicked' problems (Rittel & Webber 1973). Adapting a phrase from the narrator in Rushdie's *Haroun and the Sea of Stories* (1990), we may describe them as P2C2Es: Processes Too Complex to Explain.

The World Food Problem:
A Complex Issue and One of Today's Great Challenges

One of today's great challenges is to feed the population of the world when it reaches an expected 9.7 billion in 2050 and to do this in sustainable ways, i.e. in ways that the planet is able to maintain given its carrying capacity. Since the food issue operates on different systemic levels and involves many actors and factors with intricate mutual feedback relations, it qualifies as a complex problem. It is not an isolated problem; at a higher level, it is closely connected to other issues such as climate change and the need to provide sufficient water and energy that results from the increased demands. The globalised society and the economy are intrinsically linked to the ecological system that provides the food, energy and natural resources we need to stay alive. The connection between three of Earth's main sub-systems is known as the Food-Water-Energy nexus. It is an essential cluster within the socio-ecological system, a cluster that is itself a (sub)system of the overall system of our planet.

Given the complexity of the food issue, it is not easy to reach agreement on defining the problem and its boundaries, let alone on the most appropriate solution.

When we attempt to define the problem, the first question we must ask ourselves is: how can we describe it? Is it a single food question, or is it a series of food questions? There is general agreement that the world's food problem is about supply and demand. Yet there is less consensus about whether the solution lies in increasing production, in redistributing the available food resources, in providing more equitable access to those resources, or whether we should be looking to change consumption patterns. Clearly, defining the problem and drawing boundaries around the issue is far from simple.

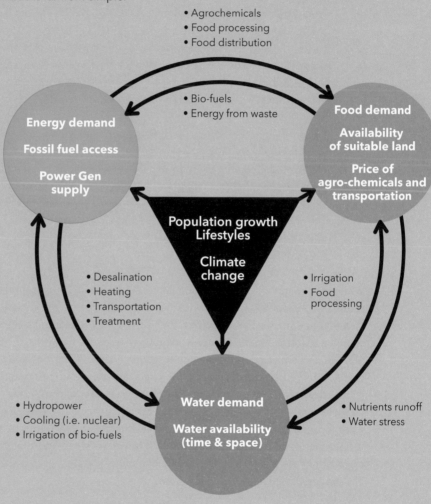

Figure 1.1 The Food-Water-Energy Nexus
Source: Smajgl et al., 2016

Let us assume, for the sake of argument, that there is agreement that the solution to the food problem lies mainly in increasing production. This still leaves us with the problem of which approach to the issue is the best and what solutions are most adequate, which remain matters of intense debate.

Some scientists claim that so-called biological solutions can increase yields while simultaneously reducing the use of pesticides. This involves using new insights into the way plants resist disease and pests in order to develop less vulnerable crops. Others propose the smart use of existing biological technologies such as Marker Assisted Selection to create stronger, cheaper and more productive ways of tackling the food problem. There are scientists who argue that all available methods, including genetic manipulation, should be used to ensure that our future food needs are met (see e.g. FAO 2004; Fresco 2006, 2012, 2015; Asafu-Adjaye et al. 2015; Visscher & Bodelier 2017). Yet other scientists propose that the latest conventional methods – for instance, precision agriculture using high-tech tools such as GPS technology and climate apps – can raise production just as effectively as controversial biotechnologies and genetic methods.

In their search for solutions, scientists often focus on technical innovations. They generally pay less attention to the possibility of social innovations, for instance adjustments to the food chain or alterations in eating habits. Yet such social changes can also have a major impact (Fresco 2012: 452-463, 478), and most scientists agree that it is more important to work out how to redistribute available food and food resources than to produce more food.

While they are not the same and should not be confused, complex problems often relate to *complex adaptive systems* (CAS) such as ecosystems, the climate, economic systems, cities or societies. In effect, the planet's socio-ecological system is a massive complex adaptive system. Complex adaptive systems are networks or collections of mutually connected, interdependent agents capable of learning and adapting to changes in their environment. Another characteristic of complex adaptive systems is their capacity for *self-organisation*: they have the power to completely change themselves by creating whole new structures and behaviours. This process takes place through numerous feedback loops (i.e. interaction and retroaction) involving various factors and/or actors. As a result, it is impossible to find straightforward explanations for the behaviour and development of a system or to predict the effects of a proposed intervention in the system.

Complex adaptive systems are non-linear and can suddenly switch from one state to another. They can even collapse, and when that happens they are in crisis. Yet they also have a certain *buffer capacity*: they are sufficiently robust to accommodate change – within limits. They are resilient. A typical characteristic of complex adaptive systems is their *emergent* properties: they develop new features by adapting from the bottom up, organising themselves through local interaction and feedback loops. Phenomena observed at the macro level cannot therefore be reduced to the features or functions of their component elements at the micro level (Midgley 2000: 39-44;

Morin 2008; Ulanowicz 2009: 115-119; Homer-Dixon 2011; DeTombe 2015: 73). They are more than the sum of their parts.

While greater connectivity within a system makes it more robust, too much connectivity and tight coupling (reducing the physical or temporal distance between elements) can threaten a system's resilience. Above a certain threshold, a tipping point is reached that can cascade domino-style, causing the whole system to collapse (ibid.: 7-9; Scheffer 2012). A system breakdown is not necessarily a bad thing – many celebrated the fall of the Berlin Wall and the downfall of the Soviet Union as welcome collapses of despised systems. Nor does the breakdown of societal or ecological systems mean by definition that they cease to operate. More often than not, it means that the conditions and composition of its components have changed dramatically: species may become extinct or resources may become depleted, for instance, or people may starve or may be oppressed. While we can expect collapsed systems to be persistent in the sense that they display continuity, they no longer provide an arena for interaction that we consider desirable (Andersson et al. 2014: 151).

Complex adaptive systems are not a recent phenomenon; they have existed for as long as people have lived on this planet. Yet over the last hundred years, the human-driven world has become increasingly complex. Certainly the modern Western societies have shown an increasing specialisation, bureaucratisation and globalisation that make life quite complicated. This is the result of the increasingly refined and ingenious social, economic and technological systems we have developed to try and solve the problems we encounter (Homer-Dixon 2011: 3-4). We live in a society that focuses on productivity and technology within a culture that celebrates quantity, efficiency, usefulness and speed. To increase the efficiency of labour and production, economic operations and tasks have been divided into subsidiary tasks. This has also happened in technology and science: problems are divided into separate elements, leading to a differentiation into specialisations and tracks. As a result, relations in social life are generalised, formalised, made anonymous and commercialised. While science, technology and the economy have grown independent of the rest of society and developed into separate systems, they still exert considerable influence in almost every sector. They are globally connected to each other but nevertheless play a pivotal role at the local level (Habermas 1981a; Beck 1986; Giddens 1991).

Ironically, the problems that our social, economic and technological systems are supposed to solve are often the results of our own actions (DeTombe 2015: 18, 32). The technologies we have designed over the past century and a half have had far-reaching consequences for life on Earth. Their impact has been so huge that theorists have introduced the term *Anthropocene* to describe the geological age that started with the Industrial Revolution, when human activity began to have a decisive impact on climate and the environment (Revkin 1992; Crutzen & Stoermer 2000). It is precisely the problems that arise at the interface of humankind and the planet – where humanity interacts with the surrounding environmental adaptive systems

– that are not merely complicated but indeed complex. Complicated problems involve multi-functional components in inclusive, nested hierarchies (implying that each hierarchy is encompassed in another hierarchy, just like Russian matryoshka dolls) in multi-layered organisational structures. Complex problems involve all this and in addition have elusive characteristics such as feedback loops, self-organisation and emergence.

As a result, complex issues appear to be a mess. They seem like problems with a disturbing unpredictability. They are inextricable, chaotic, ambiguous and full of uncertainty (Morin 2008). Scientists do not like this; they prefer orderly structures and clear connections amenable to manageable classifications, comprehensible explanations and verifiable predictions. They try to create order in these phenomena by suppressing chaos and eliminating ambiguities and uncertainties. While this strategy may seem logical and necessary, it risks pushing aside essential elements of complexity. Yet we still have to find ways to deal with this increasing and apparently growing complexity.

1.2   The State of Modern Science

1.2.1   Foundations of Modern Science

There is a saying: 'knowledge is power'. But how does knowledge make us powerful? One answer is apparently that knowledge gives us insight into underlying chains of cause and effect. This enables us to understand how phenomena work. And once we know how they work, we can manipulate situations and exercise control over our surroundings. We can organise our environment to create optimal conditions for our needs. Modern science, from its beginnings in the late seventeenth century, is designed specifically for this purpose. Particularly in the natural sciences, tremendous progress has been made by exploiting the insights we have gained.

> ### Knowledge is Power:
> ### Food Production, Modern Science and World Domination
> Wild plants developed into edible crops as the result of a partly deliberate but at first probably largely unarticulated selection process by farmers. In fact, it is only in the last 11,000 years that people turned to what we now call food production. Over thousands of years, farmers selected individual plants based on perceptible qualities such as size and taste as well as imperceptible aspects such as seed dispersal mechanisms, germination inhibition and reproductive biology. Much of the transformation of wild plants into edible crops was also achieved by plants selecting themselves. Our understanding of how this works received a huge boost in 1859, when Darwin published the idea of a general mechanism of natural selection, a core aspect of his theory of evolution. Until then, we had no generally accessible record of important determinants of plant fertility such as causes of disease, the role of nutrients and genetic processes.

The rise of modern science in the late seventeenth century gave us greater insight into the underlying processes of crop growth and food production. During the Industrial Revolution, farmers began to till, fertilise, irrigate and weed their land more effectively. The changed environment improved conditions for growth and stimulated the further domestication of plants. In around 150 years, accelerating scientific research into the genetic composition of crops has given us an ever-greater understanding of how to improve the production of seeds.

Similarly, wild animals were transformed into farm and domestic animals. Here, too, modern science has increased our knowledge about the wildlife that lives (or once lived) around us, and with that our ability to tame these creatures and use some of them for our own purposes.

In *Guns, Germs, and Steel: The Fates of Human Societies* (1999), Jared Diamond explains how favourable circumstances for food production and animal domestication allowed Eurasian societies to acquire a leading position in the modern world order. He argues that Eurasian societies used their advantage to conquer the world and gain more wealth and power than Native Americans, Africans and Aboriginal Australians. Their competitive advantages enabled Eurasians to develop more sophisticated institutions and advanced technologies, which strengthened their position even further. In that sense, there is a close connection between food production enhanced by modern science and the wealth and power of particular societies in the global world order.

Modern science is founded on *rationalism* and *empiricism*. Rationalists hold that reason is the main – if not only – source of knowledge. By contrast, empiricists maintain that sense experience is the primary source of knowledge. Proponents of these opposing views about knowledge were locked in a debate for many years until Immanuel Kant brought the two together in his *Critique of Pure Reason* (1781). He stated that knowledge can only be acquired by combining reason and experience. Knowledge requires experience, otherwise it has no content. Without experience, it is unsubstantiated and empty, for there is no such thing as predetermined, innate knowledge. Knowledge requires reason too, since humans would never be able to order and interpret the huge range of experiences without it.

In the mid-twentieth century, these insights culminated in the idea that valid knowledge requires both careful, unprejudiced observation and thorough systematic theorising. The modern concept of science thus came to rest on basically two foundations:

1   Knowledge can only legitimately be inferred by logically valid reasoning.
2   Knowledge must be inferred from empirical observations, i.e. experiential facts.

1.2.2   First Foundation: Valid, Logical Inference

Modern scientific thinking is founded primarily on the principle that valid theories must be based on purely logical reasoning. Researchers must abide by the rules of valid inference formulated in logic. And by implication, scientific theories must be free of logical contradiction. A typical example of logically valid reasoning is a *syllogism*. Take, for example, a universal affirmative or *modus ponens*, a conclusive proposition. This type of syllogism is based on *deduction*, a top-down form of logic by which we infer the particular from the general. The statement takes the following form:

All X are Y
X

_____

Y

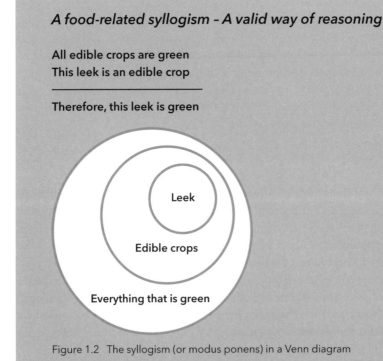

Figure 1.2   The syllogism (or modus ponens) in a Venn diagram

However, there are several issues that make deduction a less-than-ideal basis for scientific research. First, the number of generally valid rules ('All X are Y') from which to infer deductions is limited. As a result, the deductive method does not lead us very far. Neither is logically valid deduction an unfailing way to reason. It is true that in deduction, *premises* lead necessarily to the conclusion reached; if the premises above the line are true, the conclusion under the line must be true too. The difficulty, however, lies in the word 'if'. For if the premises above the line are not actually true, this can lead to errors.

But the main problem of deductive reasoning is probably that the correct application of the method usually produces nothing of note. That a leek, like other crops, is green, seems a little obvious. It would be far more interesting if we could conclude that leek, which is identified as one of a group of green vegetables, is edible. This kind of reasoning produces far more useful insights and is often employed in science. However, this method, in which premise and conclusion are reversed, is a logically invalid form of reasoning. The fallacy, known as a confirmation of the consequence, entails that we can never prove a cause by *validating* the effect.

A food-related form of invalid deductive reasoning (the confirmation of the consequence fallacy)

All edible crops are green
This leek is green
_____

Therefore this leek is an edible crop

Because of the problem related to the top-down *argumentative* process of deduction, modern science adopted another form of reasoning: *induction*. Unlike deduction, induction involves bottom-up reasoning, starting with singular observations, on the basis of which a general theory is developed. For example, we assume that since most crops we eat are green, all edible crops are green. In this way, we formulate rules based on large amounts of empirical evidence that confirm our theoretical suppositions.

A food-related induction – A formally non-valid way of reasoning

A leek is green
Cabbage is green
Lettuce is green
_____

All edible crops are green

Initially, scientists tried to establish hard facts by registering repeated observations that without exception led to the same results. These facts formed the basis of general theories and causal explanations of certain natural phenomena. The combined laws of nature were viewed as a cohesive system of scientifically proven facts about cause-and-effect connections. (The link between rationalism and empiricism is obvious here: we need empirical experiences to be able to infer general laws.)

Scientists sought out experiences and results to confirm their theoretical explanations. For researchers, the main task was to check whether the facts agreed with the expected results, or to look for facts and arguments to confirm the theory.

However, Karl Popper (1970, 1972, 1976) showed that *verification* – i.e. the confirmation of formulated hypotheses – does not provide a watertight guarantee that a theory is true. While a researcher may find a thousand swans that are all white, the thousand-and-first swan may contradict the theory that all swans are white. If the last swan observed is black, the entire theory turns out to be invalid. In our food example, we need only find a red beetroot to show that not all edible crops are green.

The lesson here is that verifying inductively inferred theories does not produce irrefutable knowledge. There is no logical scheme that enables us to conclude from the observation of a limited number of specific cases that the observed phenomenon applies in all cases. No matter how we try, a general statement can never be confirmed by a collection of supporting examples. This is known as the *induction problem*.

So we are left with this: neither deductive nor inductive reasoning are without flaws. In chapter 2, we examine how modern science has tried to find a way out of this problem by combining both ways of reasoning and by adjusting the testing procedure. But now we take a closer look at the second foundation of modern science: empirical observation.

1.2.3 Second Foundation: Empirical Observation

We have seen how scientists draw on series of observations in particular situations to formulate general theories and laws. Originally, scientists made empirical observations and recorded sensory experiences to confirm a proposed causal explanation. For researchers, the main task was to check whether the facts agreed with the anticipated results, i.e. to gather findings and arguments to help confirm the *hypothesis*. Where this involved explanations of non-observable phenomena, they tried to test these by examining observable implications of the explanatory model.

However, just as the use of logic drew criticism, the empirical fact-checking method also came under close scrutiny. It was particularly the belief that direct observation of the world is possible that drew many objections (Quine 1953; Popper 1970, 1972, 1976). This belief rests on the assumption that thorough, unbiased observers can directly acquire facts through sensory perception independent of any theories they may hold. In other words, we assume that what we perceive is determined entirely by the nature of the observed phenomenon. This is based on the idea that nature speaks for itself and that empirical facts are a given. But actually, our search for facts depends on the knowledge, experience and insights that we have developed earlier. To obtain accurate observations, we have to learn how to observe. Relevant aspects of observations must be distinguished from those that are not relevant, and they must be recorded as such. This is a trial-and-error process that is not without faults and sometimes includes mistakes (cf. Koningsveld 1987: 132, 199; Chalmers 1999: 1-40).

Theory-Laden Observation: Learning How to Study Plants

Since we know that plants consist of more than the parts we observe above ground, when we study plant growth we also examine the roots, which are not directly observable. We can understand why the growth of a plant stagnates if we know that it needs light, water and nutrients and we see that one of these elements is missing. We can discover what causes a plant to look weak after we realise that plant eaters, fungi, bacteria, viruses, insects and roundworms can impede its growth. And we can develop insight into the role of plants in the planet's carbon cycle once we know about photosynthesis. So before we can register useful observations and make a valuable contribution to the existing body of knowledge in the botanical field, we already need considerable knowledge of plants.

What we acknowledge as fact depends on the theory on which we base our research. Moreover, our observational capacities are limited; we can only view and measure what we perceive. In some areas, we may lack the means to observe – we only need to be reminded of the new worlds discovered since the introduction of advanced microscopes to realise our limitations. So to claim that a direct one-on-one relation exists between observation and statements based on observation would be untenable. To bring order to our observed reality, we need a conception or theory.

1.2.4 The Ideal of Unified Science

The attempt to create a standard model for science was accompanied by the aim to develop a unified science (see Neurath, Carnap & Morris 1952). This required a uniform scientific language and a uniform research method. Given that the building blocks of a theory are arranged from the general to the particular, the ideal would be to establish uniform rules, with those of physics at the top. Physics would provide the foundation for natural scientific disciplines such as biology and chemistry as well as the social sciences such as psychology.

Motivated by the success of the standard method, many scientists came to agree on the principles governing how research should be conducted. Until the 1960s, there was broad agreement in the Anglo-Saxon world regarding the basic assumptions and methods with which scientists should conduct their research.

Even the social sciences embraced the research principles developed in the natural sciences. Anticipating that their new branch of science would develop rapidly to reach the level of its older sibling, the natural sciences, social scientists adopted the standard method in many areas of social research. As a result, this method came to dominate the social sciences too. Even in the more established disciplines of the *humanities*, researchers focused on finding patterns and rules (see Bod 2012: 20-21, 424).

This agreement regarding the assumptions and methods of scientific research is referred to as the *orthodox consensus*. One of the main assumptions underpinning this consensus is that the natural sciences can and should be the model for the social sciences. This is known as the *postulate of naturalism* (Giddens 1985: 27-28). The postulate holds that there is no reason to assume that the social sciences and the humanities are dealing with a fundamentally different research object. Nor is there any need for scientists to doubt that naturalism and its accompanying concepts and methods are a useful and fruitful approach to the whole scientific enterprise. Accordingly, the principles of objectivity and generalisability are considered to be the leading criteria in every domain of scientific research.

Another assumption in the standard scientific research model is that research objects behave according to certain laws. This is called the *nomothetical postulate*. Scientists are not necessarily interested in discovering a particular cause of a certain effect for a specific phenomenon. What they really want to establish is a law that determines that a particular cause has a certain effect on all such phenomena. Ideally, independently operating factors can be viewed as specific causes and effects of generally applicable causal relations (De Boer 1980; see also Flyvbjerg 2001). Scientific statements should therefore be predictable, such as the proposition 'If we drop a stone (= mass), it will necessarily fall to the ground (due to the gravitational pull of the earth).' Or: 'Each time we reward someone for particular behaviour, that person will be guaranteed to repeat that behaviour.' The uniformity of this approach is clear: it should be possible to reason in the social sciences in the same way as in the natural sciences. And this should also apply to the humanities.

The uniformity of science is furthermore evident in the postulate of analysability. This states that it is theoretically possible to break down and study a phenomenon in distinct, independently operating factors or *constituents*. Reality is generally viewed as a system of mutually related functions that can be approached and studied as separate units. In other words, the phenomenon is approached from a *reductionist* or *atomistic* perspective. We thus often see in experimental situations that only part of a phenomenon is studied. Indeed, the smaller the constituent, the fewer the factors that need to be controlled to obtain objective results. When we want to know the overall picture or the overall effect, the various research results are 'added up' to compile a total final result, as it were.

Often, the metaphor of a machine is used to illustrate the reductionist or atomistic approach. Essentially it is based on a mechanical model that operates through causal connections that can be demonstrated in experiments (like the model of a dam: a small-scale example showing how water behaves in a full-scale situation). Models are especially useful in generating ideas for new theories or making adjustments and additions to existing theories. Visualising an insight in a model can have a powerful impact on the search for explanations of particular phenomena. Presenting a phenomenon with which we are unfamiliar in a familiar form can make it easier to find possible explanations. Yet a model's usefulness can also be its pitfall: suggestions derived from models can be deceptive. Models can mislead; we always need to test them carefully (De Boer 1980).

Technical advances have also led to a modern variation modelled on the computer. Here, instead of the nomothetical postulate, the programmability postulate applies. This is founded on the view that computer programmes can also provide a good model for explaining and predicting in science (ibid.; Ulanowicz 2009: 37). The machine analogue is increasingly making way for new images in which eco, societal or eco-social systems are depicted as an anthill or a flock of birds, reflecting the view presented in complexity thinking that patterns or structures are not the result of machine-like order but rather of organic disorder and a certain emergent messiness (McCracken 2006: 114). The new metaphors reflect the idea that a system comprises a self-organised overlay network of interconnected nests. An anthill is a network characterised by the absence of a fixed structure: nests come and go, depending on the input and output of the ants and how they react to each other. Yet perhaps these self-organising, messy systems can be programmed and modelled, not only within the domain of the natural sciences but in the social sciences and humanities as well. Below, we take a look at the role played by computer models in contemporary science and at their implications for the knowledge process.

1.2.5 Dissent against the Orthodox Consensus

While there is some consensus regarding the postulates we have discussed, this does not extend to all scientists. There are many scientists who refuse to accept that these basic principles of the standard model should or can apply universally to all scientific research or academic studies.

Opponents of the standard view have pointed out that basing research on these premises prevents researchers from doing justice to the real character of what they are researching in the social sciences and the humanities: people, their actions, artefacts and social phenomena. These critics note that it can be difficult to reduce human behaviour to separate constituent parts and that it is usually impossible to logically infer predictable actions from this. The idea that human behaviour can be analysed from such a reductionist perspective and that it proceeds according to definable laws therefore seems untenable (Morin 2008; Ulanowicz 2009). In complexity-thinking terms, it is possible to say that people are complex beings that show agency, self-organisation, non-linearity and emergent properties. As in complex systems, human agency is more than the sum of its parts and is therefore difficult to predict. Indeed, it is extremely difficult to find universally applicable rules to explain human action. Either way, these critics reject the atomistic approach in which people are treated as divisible units, passive receptors of a given and self-evident reality. They are more inclined to accept a *holistic* approach in which people are perceived as indivisible units who are thinking, feeling, active beings (De Boer 1980; see also Byrne 2014: 3).

Another problem identified by critics in relation to the nomothetical postulate and the postulate of programmability is the need for objectivity. Subjective factors that can influence a person's thinking and actions – like character, emotions, personal motives and cultural contexts – are reduced to objective elements. Surveys or interview questions are formulated to leave no room for respondents' reservations and ambivalence. For example, to calibrate responses about the value that people attach to certain public goods, the question is operationalised by measuring their 'willingness to pay'. Respondents declining to place a monetary value on public goods because they disagree with the implicit assumptions behind the predefined categories remain literally unheard; their non-responses are regarded as spoiled and are consequently ignored. Critics therefore claim that reduction inevitably leads to a loss of meaning: reliable *operationalisation* is by definition impossible.

Another issue related to the orthodox consensus is the failure to acknowledge that, unlike research objects in natural science, people are conscious beings who interact with the experimental setting. People understand what is going on, they attach meaning to what happens to them and around them, and act according to their own views and goals (Byrne 2014: 64). Applying approaches from the natural sciences to the social sciences or the humanities would therefore produce data that are of little use.

Research results from the social sciences and the humanities offer small comfort; their efforts don't amount to much compared to the successes of the natural sciences. The lack of generated general laws suggests that the principle of generalisability of knowledge rarely seems to apply in these domains.

Criticism of the standard model has led in practice to different ways of working in different academic disciplines. So it would be inappropriate to focus exclusively on this model as if it exerted complete dominance, since in the social sciences and the humanities there has never really been a main movement. In the twentieth century, other approaches developed in the social sciences and the humanities that are based on quite different foundations. In the next chapter, we examine these research approaches individually and explore their contribution to the study of complex challenges. But first, in the remainder of this chapter, we take a closer look at the inevitability of assumptions in scientific study and how best to deal with these.

1.3 The Inevitability of Philosophy

If we describe the aim of philosophy in general as the attempt to find out how we understand reality, we can say that philosophy of science narrows the investigation down to the domain of scientific knowledge production or knowledge creation. Philosophy of science is dominated by questions such as 'What is knowledge?' and 'What is truth?' When we make a claim, how do we know it is true? What is the ultimate foundation of scientific knowledge? Do we actually know anything for sure, or is our knowledge inevitably accompanied by inherent uncertainties? If so, can we ever overcome these uncertainties? In short, philosophy of science focuses on the analysis of how we acquire knowledge and how we evaluate and value that knowledge.

It is often said that philosophy is of little importance to scientists. As it consists purely of theories that are eventually replaced by other theories, some claim that philosophy has little practical value and only serves to confuse. This is not a view that is shared in this book, and we now explain why.

Philosophy's principal task is to reveal the assumptions behind specific arguments. In philosophy of science, these are the implicit assumptions on which we base our knowledge and our research methods – assumptions that often seem so obvious that we hardly notice them. Analysing assumptions and premises enables us to evaluate research critically. It helps us to assess whether the conclusions we draw from our research are correct and well-founded. In that respect, it is essential to have a background knowledge of philosophy to be a good scientist. Philosophy of science is a crucial part of any academic course precisely because it is vital in science to be able to show what a particular piece of knowledge is based on. As Collier (1994: 16) says:

> 'The alternative to philosophy is not no philosophy but bad philosophy.
> An unphilosophical person has an unconscious philosophy.'

Just because people say they make no assumptions does not mean that this is true. At most, it means they are unable or unwilling to reveal them. It is the open and critical examination of a person's own basic principles that distinguishes a genuine and conscientious scientist from a pseudo-scientist. Scientists whose assumptions are implicit or who refuse to change their position when new developments challenge these assumptions are actually basing their work on ideas more akin to beliefs. They have left the field of science and have entered the world of religion.

Below, we identify three ways in which scientists are confronted with assumptions. Before we do that, it must be pointed out that there is no way to avoid assumptions, since there is no solid irrefutable foundation for our scientific knowledge claims. Next, we examine the role of the overarching 'frames of mind' – the theoretical and methodological frameworks that groups of researchers involved in solving scientific problems take for granted. Similar frameworks also apply to philosophy of science, as the subsequent section will show. An examination of various successive frameworks demonstrates that the questions asked by scientists, and consequently the types of solutions that they bring forward, depend to a large extent on the frameworks that guide their research.

1.3.1 The Münchhausen Trilemma

At the heart of the issue is the following: there can be no science without assumptions. Science is and will always be based on assumptions. There is no way for one not to make assumptions. It is impossible to get rid of them. Any attempt to arrive at an absolutely irrefutable proof of acquired knowledge leads to a situation similar to that of Baron von Münchhausen when he tried to pull himself out of the swamp by his own hair, which is why this is known as the *Münchhausen trilemma* (see Baynes et al. 1987: 251, 302). It is called a trilemma because there are three available options to justify acquired knowledge, none of which provides absolute or irrefutable proof.

The first option is to seek an absolute, irrefutable foundation by turning to an authority. This leads to an arbitrary, dogmatic conclusion to the argument: a particular method or theory is simply declared as *the* touchstone or *the* basic foundation. Questions such as why a given argument applies are answered simply by saying: because X says so. This strategy has been rejected in modern science as untenable; we refuse to accept the idea that a statement by an authority can be true by definition. What the Church or Aristotle (the authorities of bygone eras) says is not the infallible Truth with a capital 'T'. Yet new authorities may operate in the same way. In modern science, for example, what can be observed in reality (in Dutch: *waar-neembaar*; in German: *Wahr-nehmbar*) is assumed to be equal to what is real (in Dutch: *waar-heid*; in German: *Wahr-heit*). Observed reality is now the infallible authority. So the trap of a dogmatic, authority-based conclusion still exists.

A second trap that scientists can fall into is a *logical circularity*. In a circular argument, the proof is founded on the position taken in the argument. Thus, in effect, the proof turns in a loop. The asserted position is simply repeated. Take the statement: "This action is punishable because it is against the law." In fact, the same thing is said twice but only in other words. The proof rests on a foundation or a statement that is assumed to be correct yet which itself remains to be proved. As a result, the argument rotates in a circle and can never produce a final, absolute proof. In the standard model, we can also point to a circular argument. Research efforts are focused on trying to establish a system of laws. Yet such a system of laws cannot be found without first agreeing on the facts to which these laws apply, as those facts can only be established based on and according to the laws they are supposed to support. The example of a biological law of pathogens in plants in the figure below shows how this works.

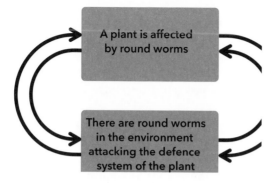

Figure 1.3 Circularity in theory construction and the search for empirical facts that serve as support for the truth of the theory

The third trap is that of *infinite regress*. This refers to the problem that the search for an ultimate proof inevitably leads from one proof to the next, ad infinitum. For example, the objectivity of a method is based on yet another method or relies on a particular philosophical assumption. So the problem effectively shifts to the legitimacy of the objectivity of the other method, or the underlying assumption. (This perceived objectivity of the method or assumption may be supported by certain authorities; in this respect, the third trap is closely connected to the first.) Here again, it is impossible to find a final, absolute proof. And this applies both to the natural sciences as well as the social sciences (see Hempel 1965: 423-424). An example may help explain how this works. Until the 1960s and 1970s, it was assumed that genes were the smallest entity in heredity. Since then, scientists have stepped forward who state that different genes cooperate in protein expression, bringing molecular biologists' dogma of 'one gene – one protein' into question. The idea that genes control heredity has also been challenged (Kwa 2014: 15-16). So what the problem of infinite regress shows is that each cause can be seen as the effect of some other underlying cause (see Popper 1957: 124), and as a result we can never obtain a final explanation. When do we stop asking why? If we ever get to a point when we say "That's just the way it is!", we fall into the trap of the arbitrary conclusion. Yet there seems little point in continuing to search for an explanation for an explanation for an explanation. So what should we do?

That is the question we are left with. It is the question we have to learn to live with, especially when it comes to scientific research into complex problems.

1.3.2 Paradigm Shifts: Tipping Points or Turning Points?

In the vast majority of fields of scientific endeavour, the work of a group of researchers is guided by a kind of disciplinary matrix of do's and don'ts made up of a collection of laws, methods, examples and commitments. It is an established way of doing research that a school of research or research movement considers normal and that represents the framework within which explanations and solutions are sought. Such a framework or system of generally agreed assumptions that are taken for granted is known as a *paradigm* (Kuhn 1962).

Each paradigm is based on at least three types of question:

1 What is our vision of reality (*ontology*): what do we assume about the world around us? What are the building blocks or constituents of reality?
2 What do we consider knowledge (*epistemology*): how do we establish valid knowledge about reality? How do we justify that knowledge, and what do we take as convincing proof?
3 What method do we use (*methodology*): given the nature of reality and the way we examine it, how do we design our research process?

In *The Structure of Scientific Revolutions* (1962), Thomas Kuhn argues that scientific disciplines develop according to a standard pattern in which different phases succeed one another. In his first formulation of this view, he describes the progress of science as a process in five stages:

1 immature science;
2 normal science;
3 crisis science;
4 scientific revolution;
5 return to normal science.

The cycle repeats in an endless loop from phase 2 to 5 and back again (see figure 1.4).

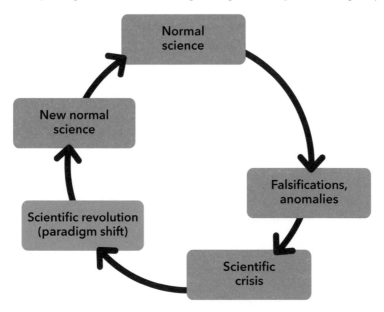

Figure 1.4 From normal science to paradigm shift
Source: Lecture Philosophy of Science Keestra 2013

At first, Kuhn described immature science as preparadigmatic. Later, however, he claimed that all forms of research take place within the contours of a paradigm. Immature science is science characterised by a paradigm that for some reason fails to establish a puzzle-solving tradition. The development from immature to mature science is therefore a development from a paradigm that is leading nowhere to a paradigm in which unresolved problems are tackled in such a way that at least some hope of a solution exists.

In mature science, there is agreement in the research community regarding the paradigm: a broad consensus exists about what is true and what is untrue. When the science world is in balance, participants are happy to work within the framework of the established dogma. According to Kuhn, most scientists spend their entire career working within an established tradition like this. In normal science, participants attempt to expand previously well-tested theories, to enhance them and to put them into practice. They solve problems and fill in the details. All seems to be under control and according to plan. Progress is made: a growing number of phenomena and systems can be viewed in terms of the essential examples of the main theories of the paradigm.

Yet no single theory explains everything, and no single paradigm solves every problem convincingly or successfully. At some point, phenomena emerge that cannot be explained by the established theories and seem to not fit in the paradigm. Calculations fail to compute, or findings contradict anticipated results. Problems that cannot be explained using established theories or that cannot be resolved using methods connected to the paradigm are identified as *anomalies*.

Now the scientific community attempts to adapt the model and to develop ways to fix the paradigm. Yet as the contradictory evidence builds up, the paradigm begins to show its faults, especially when rival researchers repeat critical experiments and develop more tests to demonstrate these faults. Defenders of the established doctrine start to feel unsure. Clearly, the damage cannot be contained. Minor repairs no longer suffice, since there is obviously something wrong with the core theories of the paradigm.

When the scientific community fails to resolve anomalies within the paradigm and cannot explain phenomena according to established theories, this leads to a crisis. Crisis science is marked by a general recognition that the established paradigm no longer functions effectively. This recognition may come from the scientific community working within the paradigmatic tradition or from outside this tradition. It is as if a certain threshold has been reached in the system – in this case, a knowledge system – and the system's resilience is pushed to the limit. As the paradigm – i.e. the established knowledge system – rocks on its foundations, new ideas, theories and methods are proposed from all sides. Now it is only a matter of time before a new alternative paradigm evolves. When one of the new paradigms emerges as a potential successor, a conflict develops between the old and the new paradigm.

Eventually a new paradigm will triumph, one that succeeds in developing a new conceptual dimension that resolves the contradictions. The space for solutions expands and the anomalies vanish. At that point, science is in *transition*, as the whole knowledge system prepares to shift to a new foundation – a situation as yet untested, in which new concepts are introduced and new rules apply, although they only become apparent over time. Gradually, a new, normal science tradition emerges.

It is not easy to determine whether scientific development shows the characteristics of a revolution or that of an evolution. Dramatic turning points in thinking, such as the advent of relativity theory or quantum theory, are often seen as revolutions. In systemic terms, a tipping point is reached, precipitating a collapse in the reigning knowledge system and heralding a new overarching approach. Some argue that this is not how science works (see e.g. Midgley 2000: 254-259; Bais 2009). They prefer the term turning point, or Kuhn's own term paradigm shift. Yet even if a paradigm shift is not really a revolution, it still involves a radical change in basic assumptions with far-reaching implications for the entire knowledge system.

1.3.3 Turns in Philosophy of Science

Ironically, Kuhn's *Structure of Scientific Revolutions* (1962) launched a paradigm shift of its own. Originally written for the *Encyclopedia of Unified Science* (Neurath et al. 1952), it actually undermined that ideal of a unified science. Kuhn's work turned out to be a Trojan horse: it discredited the attempt to find common ground or a basic foundation for all science. It showed that it was impossible to establish criteria for good science once and for all and that these actually change as science moves from one phase to the next (Toulmin 1990: 84, 2001: 22; Bradie 2006). Kuhn's publication led to a pragmatic turn in philosophy of science: rather than painting an ideal picture of science, attention now came to focus on scientific processes as they actually happen.

As in any other discipline, fundamental shifts can and do take place in philosophy of science itself. Ideas about what science is or should be and how the process of obtaining scientific knowledge evolves have changed profoundly in the last century. It is beyond the scope of this book to highlight them all. But one turn that deserves to be mentioned here is the new potential created by the introduction of the computer. Its impact, which extends beyond the physical realm and visibly affects the social sciences and the humanities as well (cf. Gilbert & Troitzsch 1999), has such far-reaching implications that some speak of a revolution in science.

The growing popularity of computer-related approaches in the arts, the humanities and social sciences has led some to suggest that a computational or digital turn may currently be taking place. David Berry claims that 'for the research and teaching disciplines within the university, the digital shift could represent the beginnings of a moment of "revolutionary science", in the Kuhnian sense of a shift in the ontology of the positive sciences [i.e. what we call modern science here] and the emergence of a constellation of new "normal science"' (Berry 2011: 10-12, 2012; see also Winsberg 2010). Were this to happen, it would imply that all disciplines have a comparable computational 'hard core', according to Berry. This in turn leads him to think that the computational turn has far wider consequences for the idea of a unification of knowledge and for the idea of the university. He goes so far as to state that computer science could form a foundation for other sciences, supporting and directing their development, issuing 'lucid directives for their inquiry'. Others (e.g. Andersson et al. 2014; Törnberg 2017) warn that the computational turn is little more than the replacement of the machine analogue by the idea of programmability, which basically is not so different. They worry that it is just another attempt to establish a unified science by reinventing naturalism all over again.

This danger might be avoided by viewing the opportunities that computational approaches offer in a broader perspective and examining these against the background of the latest developments in *systems thinking*. Systems thinking has a long history. In the social sciences, it is anchored in Talcott Parsons' *The Social System* (1951). In the natural sciences, the idea of dynamic systems theory can even be traced back to Newton's mechanics. Combined with new digital and computational possibilities, systems thinking has developed into a sophisticated approach to the world. Together, systems thinking and computation may be responsible for a more encompassing complexity shift or even a complexity revolution (Urry 2005). Given the importance of this last change in the scientific zeitgeist, we will reflect on this development more elaborately in the final section of this chapter. Before that, let us explore the need for a new, integrative approach in today's social and scientific constellation.

1.4   Future Avenues

1.4.1   The Need for an Integrative Approach

It has often been said that when we look for answers to major questions such as the issue of climate change, food or water, we cross the boundaries of mono-disciplines (e.g. US National Research Council 1999: 283; agronomist Fresco 2003: 6 and 2012: 15, 37; Platform Future Earth – Mauser et al. 2013: 421; complex systems researcher Sloot 2016). While three hundred years ago there were no separate disciplines such as mathematics, physics or political science, in recent centuries we have become increasingly discipline-oriented, and it seems that we have found ourselves stuck. We have no overarching method, strategy or framework to enable different disciplines to communicate. Meanwhile, the complex problems we are faced with are so *multidisciplinary* that they seem to be discipline-less (ibid.).

Complex systems research focuses on studying phenomena without regard to particular disciplines. It simply examines how systems interact and how they influence each other. This does not mean that disciplines have become irrelevant. There are useful aspects to disciplines, such as the deeper disciplinary knowledge and the dedicated tools and methods that disciplines use to test and validate newly acquired knowledge.

To explain and resolve complex issues, we need to connect different perspectives from different disciplines: earth sciences and ecology, for example, as well as natural scientific and social scientific disciplines in combination with the humanities (National Research Council 1999: 283; Newell 2011; Repko 2012: 84-89). In short, we need an *interdisciplinary* approach. As proposed in *An Introduction to Interdisciplinary Research* (Menken & Keestra 2016:13):

> 'Today, many phenomena and problems that we are trying to explain and solve indeed 'cut across' the traditional boundaries of academic disciplines. Modern technological developments and globalisation add to the complexity of problems, and in response we are becoming increasingly aware that an integrated approach is necessary. Healthcare, climate change, food security, energy, financial markets, and quality of life are but a few examples of subjects that drive scientists to 'cross borders' and engage with experts from multiple fields to find solutions. In short, complex questions and problems necessitate an interdisciplinary approach to research.'

Given the complex issues facing humanity and the planet, there is a clear need for those working in the natural sciences, social sciences and the humanities to join forces to conduct integrated research. Their views of reality may have diverging implications for this integration, yet if we are to take advantage of the diverse approaches, we need to find a way to merge their different perspectives (Maiteny & Ison 2000; Urry 2005: 8; Mauser et al. 2013: 423; World Economic Forum 2016: 3, 25, 32).

At the same time, we also need to combine the perspectives of the various participants involved in the research process (Mauser et al. 2013: 421). When this also involves non-academic stakeholders such as politicians, policymakers, NGOs and citizens, we call it a *transdisciplinary* approach.

Hopefully, examining the philosophy of the natural sciences, social sciences and the humanities can cast more light on what these different conceptions can contribute to resolving our knowledge questions. Highlighting the assumptions behind the different research approaches and their accompanying methods may help to develop the competence to decide which approach is appropriate for a particular part of some problem. That way, scientists can either choose a monodisciplinary, multidisciplinary, interdisciplinary or transdisciplinary approach to today's complex problems. Their contributions may range from conducting basic research or supplying scientifically founded knowledge and developing policy strategies to addressing the ethical issues that inevitably accompany the great challenges of today.

1.4.2 Complexity Thinking: A New Paradigm in Science?

The digital or computational turn and the insights proposed by the new generation of interdisciplinary experts can be taken to be characteristics of a broader complexity turn or even a complexity revolution that may be taking place today (Shackley et al. 1996: 202, 204; Urry 2005). Some philosophers of science claim that the advent of complexity thinking is part of a paradigm shift that involves every aspect of science (Nowotny 2016: 42). Others argue that such a paradigm shift is vital for scientific development to proceed to the next stage (Sterling 2004, 2007, 2009, 2011; Byrne 2005; Morin 2008). They believe that, if we want to shed the one-dimensional perspective dominating our current outlook on reality, we must disengage from the paradigm of simplification that has distorted our thinking.

Edgar Morin's analysis shows that the selection process steering the acquisition of knowledge often involves either differentiation or unification. It divides elements into established hierarchies or organises them around a core of principal concepts. Yet taken alone, both operations are too simple. Another simplification often employed is the reduction of the biological to the physical and the human to the biological. So we see that the main scientific paradigm is still dominated by the principles of division, reduction and abstraction. Morin argues that these have pushed us away from reality rather than drawing us closer to it. Because hyperspecialisation has divided the complex fabric of reality into pieces, complex entities have been allowed to slip through our fingers. As a result, according to Morin, we have acquired blind intelligence: an intelligence unaware of unity and totality. Because we have isolated phenomena from their surroundings, we no longer see the bigger picture, although we ourselves are at least partially responsible for isolating them in the first place.

The difficulty with complexity thinking is that it has to do justice to the intricate connections and continuous feedback loops between the phenomena. Moreover, it has to deal with the messiness, the uncertainties and even the contradictions accompanying this process (ibid.; see also Nowotny 2016). Fortunately, we have several conceptual tools at our disposal that can help us develop the new complexity paradigm. What we need to do, Morin explains, is abandon the one-dimensional paradigm in which distinction goes hand in hand with reduction and replace it with a paradigm in which distinction is followed by conjunction or unification. For it is possible to distinguish without permanently separating the constituent parts. Ultimately, the aim is to connect elements without creating an oversimplified, unidirectional hierarchy and without reducing the parts to simpler entities.

It is difficult to know whether the complexity turn has already launched a paradigm shift or whether this has yet to start (or if indeed it will). Can we at least regard it as an inspiring virtual paradigm (cf. Yolles 1996: 558) that offers us a new approach to tackle the complex problems at the interface of humanity and the planet? Or is it just another case of disciplinary imperialism, an unavailing attempt to bring the social sciences back into the realm of the natural sciences to unify all the sciences? It remains to be seen.

Closely related to a critical assessment of this present shift is the discussion about whether the new approach should be referred to as complexity science. Cilliers (2004: 136) remarks that this depends on what is meant by science. Taking the formal definition, which demands formal notation in symbols and syntax, complexity thinking cannot be regarded as a science. Yet our growing knowledge of complex systems is undermining such a strict conception of science, Cilliers explains. It forces us to consider strategies from both the social and the natural sciences, to incorporate both *narrative*-based approaches and computational approaches – not to see which is better but to help explore the advantages and limitations of each.

So the definition of science may require a reassessment now that complexity thinking is reaching maturity. To avoid the possible misconception of associating science with natural science, here we refer to complexity thinking. We regard complexity thinking indeed as a virtual paradigm and we explore the conditions required to maximise its potential. To be able to do that, we begin the next chapter by taking stock of contemporary approaches in science. We assess their characteristics by examining their strengths and weaknesses relative to their assumptions, and we attempt to show their role and importance in the study of complex problems.

Questions:

■ What are 'wicked' problems – what are their characteristics?

■ Why is complexity a key concept in research on the challenges confronting today's society?

■ Rationalism and empiricism are apparently diametrically opposed principles of knowledge acquisition. What do these principles entail?

■ Which principles of scientific research have been criticised by opponents of the ideal of unified science?

■ What do you think of the standard scientific method for examining the domains of the social sciences and the humanities? Do you think social reality requires different research methods, or can it be studied in the same way as the natural sciences?

■ Many often remark that philosophy is not important to scientists? Why is that?

■ Why does Collier assert that a good researcher must have a sound basis in philosophy?

■ What traps lie in wait for those attempting to prove scientific theories?

■ What is a paradigm?

■ What kind of questions does ontology discuss?

■ What kind of questions does epistemology discuss?

- What is an anomaly?

- How do scientific revolutions occur, in Kuhn's view?

- Have you discovered the different paradigms used by your teachers, for example, the different paradigms used by teachers of natural scientific and social scientific disciplines or those with a humanities background?

- Why is an interdisciplinary approach vital for solving complex, 'wicked' problems?

- What is complexity thinking about?

- What do you think: is a paradigm shift needed in science, or is it unnecessary?

2 Contemporary Approaches

In this chapter we examine three approaches to scientific knowledge production that have evolved over the last hundred years. Two of these approaches developed over a period of over half a century, with roots that date back even further. One is more recent. In some respects they are crucially different, with assumptions that vary profoundly. While this can lead to tensions, we can see how the various approaches shed light on different yet equally important aspects of scientific endeavour in the world, particularly with regard to the study of complex problems at the interface of humanity and the planet.

We start with an exploration of how the standard scientific research method has evolved since the mid-twentieth century. As chapter 1 has shown, this valuable method has helped provide insights into the structures of our world and explanations of key phenomena around us. But we also address possible objections to the proposed scientific procedure.

In the social sciences and the humanities, the diverging points of view regarding the basic assumptions underlying the 'standard' model have led to an alternative approach that is focused on understanding rather than explaining. Instead of aiming to discover truth through logical inference and empirical observation, the focus is on the role of interpretation in the *construction* of knowledge. This alternative throws new light on the scientific enterprise, although it also has its own problems.

Under the banner of complexity thinking, we are currently witnessing the rise of another approach to science as systems thinking evolves, using computer models and simulations. It resembles the traditional way of dealing with reality, since explanation is still seen as an important foundation of knowledge. But it also transcends the standard scientific method by expanding investigations beyond reality as it is now and extending into the future to say something about reality as it potentially can be.

After reviewing all three approaches, we can conclude that there is more than one way to conduct scientific research. If this is a problem, then it is a luxury problem, since we probably need every available approach to obtain the range and depth of insight needed to tackle the complex issues of today.

2.1 The Traditional Standard Research Model

Initial optimism regarding the modern scientific knowledge ideal was later tempered somewhat as critics re-examined the logical foundation and the function of empirical evidence. Later variations of the concept of knowledge acquisition attempted to meet these criticisms while retaining the basic assumptions. The result is a method in which the deductive approach of rationalism is combined with the inductive, more experimental approach of empiricism. Scientists also attempted to refine the underlying logic as well as the interaction between inductive and deductive lines of reasoning in the knowledge process. Adding a statistical and computational component has completed the evolution of the standard research model, for now at least. Below, we review how this process is given shape in actual research practice.

2.1.1 The Empirical Cycle

In traditional standard research, various phases can be distinguished.

First, observations are made regarding the research matter. This may be a natural process, a social phenomenon or a human artefact. Observational activities may take any form.

Next, ideas are generated involving potentially adequate explanations for the findings of the observational phase. These provide the foundation for the theory. Theory construction is based on inductive reasoning: observations on a particular level lead to inferences pertaining to the general level. When repeated observation points to the conclusion that every swan is white (Popper's famous example) or that all edible crops are green (our food example), this may suggest a theoretical proposition: "All swans are white" or perhaps: "All edible crops are green".

When sufficient empirical evidence has been found to formulate a theory, predictions can be made based on the explanatory model. Predictions are based on deductive reasoning: the general theory forms the foundation from which specific expectations are inferred. Our two theories may lead to the following predictions: "Every single swan we encounter will be white" or "Every single edible crop we find will be green".

In the subsequent phase of research, expectations are translated into well-defined hypotheses. The hypotheses are tested in an experimental setting; they are exposed to empirical reality to see whether they hold true.

Finally, the research results are evaluated. They may also be treated as new observations. This way, the evaluations of current research results may serve as the basis of theory construction for future research. Since it is possible to build one research project on another and continually work on the further development of a theory, this method is called the *empirical cycle* (see figure 2.1).

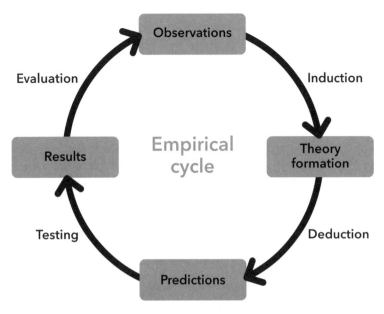

Figure 2.1   The Empirical Cycle

The Search for Causal Explanations: Mendel's Laws

For years Gregor Mendel, often known as the father of genetics, grew peas in the yard at the monastery where he lived and worked. He recorded the growth of each of his (numbered) plants meticulously, and listed which plant fertilised the other (he pollinated the pistils with a brush and cut away the stamens). Based on numerous observations, he formulated a theory about how characteristics are passed on by inheritance and breeding. He reasoned by induction, arguing from the particular to the general.

Mendel assumed that the characteristics of reproductive cells were permanent units (factors), and the combination of two of these units determined the characteristics of the plant produced by their union. When the reproductive cells meet, they randomly combine two characteristics (alleles): these may be the same, which results in a homozygote, or different, which results in a heterozygote. One of the hereditary factors may dominate the other, in which case this feature always prevails when it is one of the inherited factors (see figure 2.2).

Mendel used the generalisations he formulated to deduce certain assumptions, on the basis of which he formed explicit predictions. One hypothesis, for example, stated that cross-breeding two different plants would result in a heterozygote. Additional experiments would show whether the anticipated characteristics were indeed passed on to the next generations. When this actually happened, Mendel took it as a confirmation of his theory. The regular patterns of heredity and breeding which he recorded in the mid-nineteenth century are still known as Mendel's Laws.

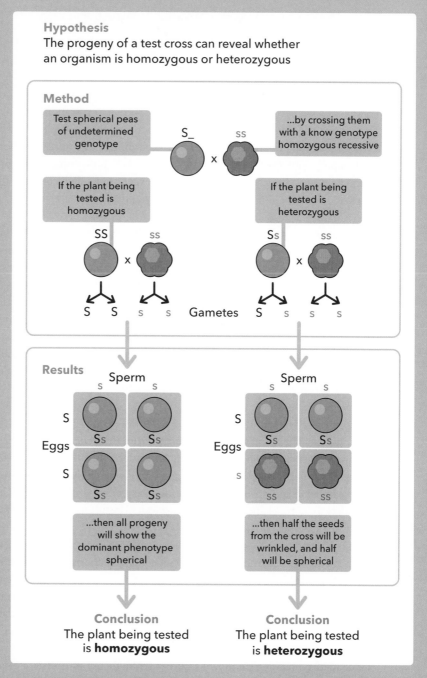

Hypothesis
The progeny of a test cross can reveal whether an organism is homozygous or heterozygous

Method

Test spherical peas of undetermined genotype

...by crossing them with a know genotype homozygous recessive

S_   ss
x

If the plant being tested is homozygous

If the plant being tested is heterozygous

SS   ss
x

Ss   ss
x

Gametes

S S   s s

S s   s s

Results

Sperm
s   s

Sperm
s   s

S

Eggs

S

Ss   Ss

Ss   Ss

S

Eggs

s

Ss   Ss

ss   ss

...then all progeny will show the dominant phenotype spherical

...then half the seeds from the cross will be wrinkled, and half will be spherical

Conclusion
The plant being tested is **homozygous**

Conclusion
The plant being tested is **heterozygous**

Figure 2.2   A theory as explanatory model
Source Sadava et al. 2014: 238

Yet it was not until the early twentieth century that Hugo de Vries (with Carl Correns and Erich von Tschermak) rediscovered Mendel's work and applied his laws of heredity to agriculture. Their application made possible a huge increase in global food production.

Originally, when scientists tested a theory empirically, they would try to confirm it. However, as we have seen, from a logical perspective, verification is problematic. This led Popper (1963, 1968) to suggest abandoning the principle of verification and trade it in for the principle of *falsification*. According to this principle, the aim is to refute a proposition with empirical facts. Instead of confirming a claim or a theory, researchers must look for ways to show that expectations deduced from the theory are false.

At first glance, this may seem odd or illogical, but think again about the swans. If there is no way to confirm a general statement such as "All swans are white" with certainty, the only way forward is to maintain the claim until it can be refuted. Instead of looking for thousands of white swans, it makes more sense to look for a swan that is not white. One non-white swan is sufficient to refute the claim that all swans are white. But as long as no swan is found that is not white, the theory of the whiteness of swans remains *plausible*. The greater the number of critical tests the theory withstands, the more plausible it is. This applies equally to our food example: as long as no edible crops are found that are not green, it is possible that all edible crops are green.

2.1.2 The Deductive-Nomological Explanatory Model

To emphasise that scientists should proceed from theoretical hypotheses rather than pure observation, Popper called his philosophy *critical rationalism*. As the philosophical debate continued, the original modern scientific ideas were adapted and expanded to accommodate the central ideas behind critical rationalism. This led ultimately to Hempel and Oppenheim's so-called *deductive-nomological model* (D-N model, see Hempel 1965).

In this model, particular predictions about actual situations are deduced from a general theory. Hence the term deductive in the name of the model. The term nomological indicates that this deduction is based on general rules (nomos = law), or at least on generally applicable regular patterns, which led to the formulation of an overarching theory.

The deductive-nomological model assumes that a theory comprises a coherent system of propositions that attempt to explain or predict certain phenomena ('t Hart et al. 2001: 140; Pradeu 2014: 22). What is key to this model is the ability to arrange the elements (propositions) of a theory in a specific order from the general to the particular. This enables scientists to formulate predictions through logical deduction from general statements and basic principles that can be tested empirically (see figures 2.3 and 2.4).

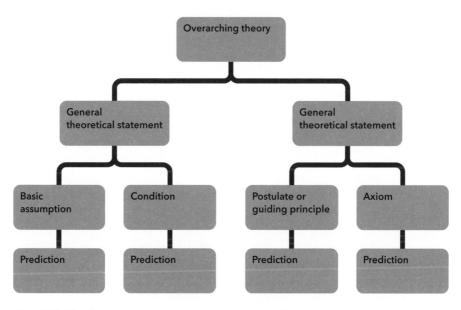

Figure 2.3   The deductive-nomological explanatory model

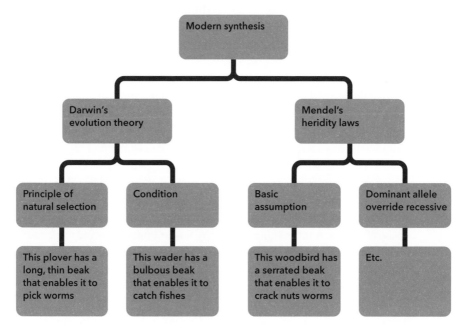

Figure 2.4   The deductive-nomological explanatory model applied in biology

While the model's name does not refer to the inductive aspect, induction still clearly plays a part in the first phase of the deductive-nomological research process (see

Hempel 1965: 425). When scientists have recorded sufficient observations and formulated some basic assumptions, they inductively develop a theory. In other words: observations at the level of particular phenomena are explained by a theory at a general level. From the observation of a large number of birds with long beaks on grassland, it may be inferred that all birds on grassland have long beaks, possibly because it is easier to find food on grassland with a long beak (see figure 2.4). This may lead to the theory that the habitat where birds live has an effect on the size of their beaks. In the next phase, expectations and predictions are deduced from the theory and then tested empirically.

The shift from verifiability to falsifiability led researchers to draw up not just one but two hypotheses: a *null hypothesis* and an *alternative hypothesis*. A null hypothesis posits that no effect will result from a particular cause. The alternative hypothesis, on the contrary, posits that certain effects expected in relation to a given theory will occur. If the null hypothesis is refuted, the alternative hypothesis may be assumed to be true. In our example, the actual hypothesis is that on grassland only birds with long beaks can be found (null hypothesis: the habitat where birds live has no effect on the size of their beaks, so on grassland birds with all types of beaks can be found). Researchers now have to look for an example to refute the hypothesis, i.e., they have to try and find a bird without a long beak on grassland. If this example is not found, the theory is not falsified, and we can say that the alternative hypothesis applies. It can be stated provisionally that it is correct that a bird's habitat affects the size of their beak. While it looks slightly different when verifiability is replaced by falsifiability, the dynamic between induction and deduction – a characteristic of the empirical cycle – basically remains unchanged (see figure 2.1).

A scientist's task is therefore to subject hypotheses to the most stringent tests. These may take the form of experiments or standardised questionnaires. Hypotheses that withstand the critical test can be provisionally accepted as true (Popper 1961, 1963, 1968). If the verification process is replaced by falsification, all scientific knowledge must then be considered temporary knowledge. We assume that knowledge is valid until it becomes untenable. This reduces science to a collection of educated guesses that we try to refute. If an attempt to falsify the hypothesis fails, then our educated guess was correct. If it succeeds, then the hypotheses founded on the theory are clearly false and science has been spared from further mistakes by a timely refutation. A proposition or theory gains in importance and value the more it survives ever stricter tests. Yet the falsifiability principle can never lead to certainty regarding a proposition or theory; this form of logic can only provide a probability model to test scientific statements.

In this way, the critical rationalist heritage eventually led to the deductive-nomological model. Yet critical rationalism was not spared criticism either, as the following shows.

2.1.3 Critical Rationalism Critically Assessed

One of the criticisms of critical rationalism is that the underlying problem remains the same, because a refutation basically depends for proof on the same method of testing theories against independent fact (Habermas 1970: 27-28; Giddens 1976: 140). Refutation is essentially nothing but disproof based on observation (Kuhn 1970: 15). In the end, this method depends on the same old idea that truth or falsehood can be established through observation. Yet it is not unreasonable to imagine that a swan may be identified as black (or non-white) although it is really a white swan covered in oil. So Popper's falsifiability principle does not resolve the problem of the second foundation of the standard method: the impossibility of observing without some kind of frame of reference, without a theoretical perspective.

We would have hoped that by now a solution would have been found to the logical dilemma regarding the first foundation, the induction problem. Popper's falsifiability principle is not based on the (formally invalid, uncertain) inductive method of reasoning but on the (formally valid, certain) deductive method of reasoning. Unfortunately, here too the solution is not watertight. First, it contains a self-referential *paradox*: it is impossible to falsify the claim that falsifiability provides the best criterion for scientific knowledge. Moreover, refutations are also based on the inductive assumption that they apply to different situations and times (Hollis 1994: 76). Thus we extrapolate from our finding that the black swan is not a one-off and we assume that similar observations will be made in the future. Yet it remains to be seen whether that is actually the case.

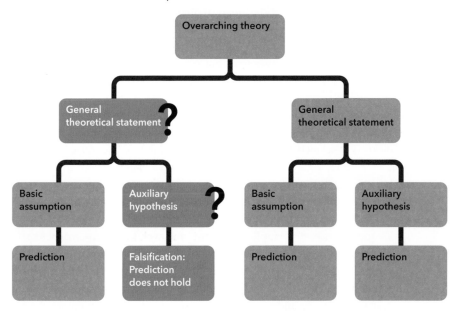

Figure 2.5 An auxiliary hypothesis can be rejected to maintain a theory after it is refuted

Philosophers Willem van Orman Quine (1953) and Pierre Duhem took this a step further and argued that science is little more than a pyramid of countless unspoken assumptions that combine to form something people refer to as 'scientifically proven'. Yet since each separate assumption is itself part of that same scientific pyramid and only hypothetical, no irrefutable evidence can ever exist. It is impossible to test an independent hypothesis because scientific theories can only be evaluated in relation to the whole interdependent structure. This problem, which has become known as the Duhem-Quine thesis, is also called *confirmation holism*. Research may involve so many factors that a simple connection between factor X and phenomenon Y is impossible to find. Since countless factors impact on countless situations, and most of these factors are not examined in actual research, it is possible in theory to prove or disprove any connection. A hypothesis is part of a network of hypotheses (see figure 2.5), and it is unclear exactly which hypothesis is being tested in a so-called crucial experiment or ultimate test.

If an observation results in a refutation, it is difficult or perhaps even impossible to establish the extent to which the cause lies in the hypothesis in question or in the assumptions that serve as auxiliary hypotheses to support expectations expressed in the tested hypothesis. Therefore, hypotheses that are unrefuted and supposedly plausible are not in themselves convincing proof.

Figure 2.6 shows the use of falsifiability in the context of Milankovitch's theory. These refutations do not lead to a complete rejection of the theory; they result in adjustments to the underlying assumptions. In this case, the alternative explanations appear plausible, and it seems sufficiently reasonable to assume that Milankovitch's theory may be justifiably held to apply. It offers a powerful, inclusive explanation for many phenomena.

Yet we should remain attentive to ad hoc rescue missions that have more to do with researchers being reluctant to give up a favourite theory than with genuine science.

We now have a reasonably sophisticated description of the standard view of science. We have traced the development of the deductive-nomologic model that prevails in many areas of science. It has brought many good things (see *'Knowledge is Power: Food Production, Modern Science and World Domination'* in the previous chapter). But we have also seen the objections raised to many of the assumptions on which the standard research model rests. Indeed, to call it standard is misleading. It is not the only approach employed in science. In addition to the deductive-nomological model, the natural sciences also employ taxonomic and evolutionary approaches that have been of great value (see Kwa 2005, 2011, 2014). The first helps create a sense of order in our knowledge of the world, which in turn leads to new questions and insights. The latter has increased our understanding of how organisms and processes in the natural world develop and has also generated insights into developments explored in the social sciences. However, rather than examine these alternative approaches, we will direct our attention to the domains of the social sciences and the humanities now and see what alternative paradigm has been brought forward there.

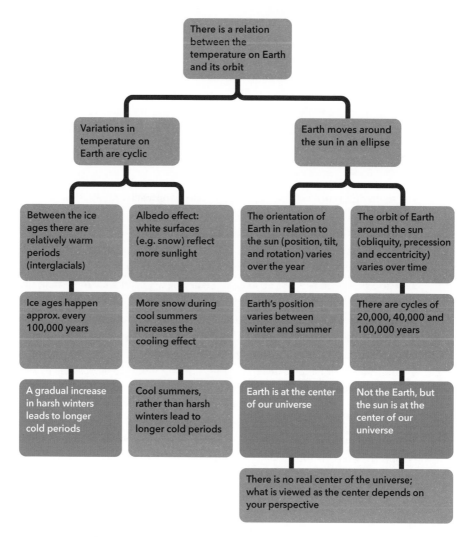

Figure 2.6 Falsifications in Milankovitch's theory

2.2   Interpretivism as an Alternative Paradigm

Influenced in part by the philosophical dispute about the orthodox consensus described in the previous chapter, other approaches have developed alongside the method that dominates the natural sciences. These alternatives place more emphasis on the importance of interpretation in understanding reality. Referred to collectively as *interpretivism*, they form an overarching approach, a paradigm in their own right. Interpretivism is a vision of science based on fundamentally different *presuppositions* than those of its counterpart, critical rationalism. Yet whether this really is an alternative paradigm is open to question, as a closer look reveals.

2.2.1 The Hermeneutic Circle

Perhaps the main reason for using an alternative paradigm to study matters in the social sciences is that people and the products and processes they create are essentially different from the phenomena studied in the natural sciences. Another approach is needed to take these differences into account. An interpretive approach does not focus on explaining the world objectively or mechanically. It aims to understand the world and people's actions in that world from the perspective of the actors themselves: how do they view the world, and what do they think about particular issues?

The main difference between the interpretative approach and the standard scientific method is the shift from explaining to understanding human behaviour (in German: *Verstehen*; Dilthey 1881). Instead of studying the external forces that motivate people to act, the focus is on an empathetic understanding of human actions driven by internal motives and personal intentions. Understanding is needed to comprehend the reasons behind people's actions, since people are ostensibly more than the passive playthings of external forces. In contrast to the nomothetic approach (i.e. finding general laws), an interpretative approach is *ideographic* and takes account of the unique and special character of human actions and artefacts. While explanation takes the form of reasoning 'from the outside inwards' (applying the general rules to the individual's situation), understanding is thought only to be possible by reasoning 'from the inside out' (trying to discover what intentions and motives drive people's actions).

To interpret other people's actions accurately, it is necessary to be able to adopt their perspective. Yet it is impossible to change perspective completely, for while we can empathise to some extent, we can never totally discard our own perspective. We can try to merge our horizon with another person's horizon as much as possible (Gadamer 1960). This is the only way to reconstruct the meaning that other people attach to their thoughts and actions. To comprehend this meaning, we need to place the text – not literally what people write but also what they think and do – in context. The context is up to the researcher to determine, from the other person's perspective, of course. Here again we see the importance of boundary work in scientific research: to interpret thoughts and actions within a specific context, it is necessary to specify this context in relation to the broader horizon of thoughts and actions.

The interpretative process appears to follow a cyclical pattern, a *hermeneutic* circle in which greater understanding is reached by continually (re)interpreting. While this resembles the process of the empirical cycle, it is based on the researcher's interpretation of what is observed rather than the empirical observation itself.

In addition to induction and deduction, the interpretative cycle also involves *abduction* or *retroduction*. Retroduction is a method of reasoning based on the principle of concomitance. The researcher tries to find the simplest and most likely explanation by pointing to co-occurrences or striking similarities in behavioural patterns between two comparable situations. As with induction, while it is not based on valid logical inference, it is a method of reasoning that we rely on when interpreting people's thoughts and actions. Choosing a hypothesis or explanation is like the choice a researcher makes about the context: the one that seems to suit the data best is the one that appears the most logical.

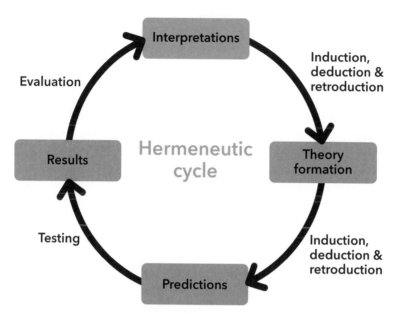

Figure 2.7   The hermeneutic circle

2.2.2   Construction and Deconstruction

Inherent to interpretivism is the recognition that people play an active role in creating the real world. Interpretivism therefore goes hand in hand with *constructivism*, i.e. the idea that phenomena are actively produced and continuously reviewed and adjusted through human interaction (Berger & Luckmann 1967; Bryman 2004: 17). This seems logical for phenomena in a social setting but less applicable to the physical world. Yet we should not underestimate the way people help shape the material world. Human understandings of nature and social adaptations to nature are profoundly interlinked. Indeed, they are so profound that Jasanoff (2004: 2; 2012: 56; 2017: 14) speaks of the co-production of the world and our knowledge of it. The way in which we know and represent the world – both the natural and the social world – are inseparable from the ways in which we choose to live in it. The two are constantly interacting, for one because the subject (the observer) has an effect on the object (the observed environment) through the definitions, classifications, representations and visualisations that are used to make sense of it all (Ochs et al. 1994).

Consider the effect of words like 'atom' or 'gene' or 'DNA', for example. The introduction of these terms has resulted in new images and theories that have dramatically changed how we perceive the world. And to give an example on a much larger scale, our idea of the climate system has only recently entered our vocabulary. Before the global weather data network was expanded in the 1970s and 1980s and became the backbone for creating the first computer models of the global atmosphere, we thought about climate merely in the restricted sense of 'a thirty-year aggregate of local weather patterns'. Thus, scientists working with these so-called general circulation models helped construct the very idea of the Earth's climate system (Miller 2004; 2017: 291). So we see that new definitions and formulations of phenomena transform the way we see the world. Our vocabulary determines to a large extent what we recognise and acknowledge as real.

Interpretivism in the Natural Domain: Active Communication by Plants

In standard biology books, DNA – the genetic material of organisms – is depicted as some sort of machine that automatically displays its preprogrammed properties. Critics have argued that this static, mechanistic presentation belies the complex way in which organisms actually function (e.g. Nesse 2013: 337-340). In their view, organisms consist of components with poorly defined boundaries, with various functions and connections linking to other elements in the environment. Rather than being predetermined, they develop in close interaction with their environment. Thus genes do more than roll out preprogrammed properties; they actively pick up signals from the environment which they either ignore or act on.

For many years, it was generally assumed that the emission of volatile organic compounds from flower petals into the air was a passive process. However, an international group of researchers recently discovered that this diffusion of molecules by flowers from their cells to attract pollinators such as bees and moths is not passive; they actively shuttle these across the plasma membrane (Adebesin et al. 2017). By representing the process in a model, they showed that the volatile compounds had to pile up in high concentrations in the cell membrane to be able to diffuse. Since this caused considerable damage to the membrane and the cell, it seemed unlikely that the passage of volatile compounds across the plasma membrane was a passive process. The finding that the process relies on active transport implies that plants can regulate the emission of volatile organic compounds and so also their interaction with neighbouring plants, insects, and benign and pathogenic microbes in their environment.

Knowledge of the mechanism underlying the emission of volatile organic compounds, and particularly of the genes responsible, can enhance our food and fruit production. This knowledge can be used to ensure that the relevant properties are maintained during selective cultivation processes.

It can also be used to improve the protection of crops against insects, since this also relies on volatile organic compounds.

The study appears to provide proof of a biologically mediated mechanism of volatile emission (ibid.: 1386). In biosemiotics (e.g., Hoffmeyer 2008; Barbieri 2008), this type of interaction with the environment may be termed communication. Adherents argue that organisms in the natural world display interpretivist tendencies comparable to those of human beings.

In the interpretivist approach, scientists are particularly attentive to the way in which experienced subjects construe their world. They acknowledge that all manners to frame and interpret the world are human constructions that are imbedded in social interaction by way of language. Knowledge, including knowledge of that which we have come to regard as 'nature', is always mediated through language (De Boer, 1988: 242; Latour 1994, 2017). Reality comes to life through language, through the stories we tell about ourselves and our experiences in the world (*narratives*) and the way we discuss and communicate matters of concern (*discourses*). Open interviews, qualitatively focused surveys, standardised or participant observations, *ethnomethodology*, studies about framing, discourse analysis and interpretative policy analysis are forms of research regularly used in the interpretative approach.

Postmodernist thinkers attach particular importance to discourse analysis. They point out that everyday means of communication allow us to express ourselves and relate our experiences. Yet they can also constrain us within existing power structures, forcing us to adopt certain norms and identities (see Foucault 1976, 1979, 1980; Lyotard 1984). The postmodernist outlook has helped us realise how the way an issue is framed can determine whether and how an issue is studied. It makes a huge difference whether climate change is called a serious problem or a hoax. By emphasising discourses and language games as a foundation of power, postmodernists show that there is another way to interpret the adage that 'knowledge is power'.

A favourite postmodernist method is *deconstruction*: the dismantling of forms and frames of communication that are thought to be forceful in order to free ourselves from inhibiting relations. An example of discourse analysis and deconstruction is the criticism of consumerism by those who portray it as an identity framework telling us to want everything. This is a framework we should abandon, they argue. Instead, we should develop another attitude, a mind-set that is more in line with a sharing economy in which people rent products only when they need them, so they are no longer slaves to their possessions, owned by what they have. This analysis of how the needs we take for granted are produced and enhanced by the traditional everyday discourse of society can set in motion a critical review of the current state of affairs.

While some find the postmodernist attitude inspiring, others consider it a dead-end street. When taken to the extreme, how can any knowledge ever serve as a foundation for coherent and progressive social action? What is the point of deconstruction if we merely exchange one restrictive framework for another? Since there is no way to establish a better or best framework, we are forced into an endless cycle of deconstruction (McCarthy 1991) – not an attractive prospect for scientific researchers.

Clearly, the way to approach social reality differs significantly from the way to study the natural world. That does not mean, however, that understanding – the disclosure of meaning – is not methodical. While it may not be possible to reduce understanding to a series of principles (like the foundations of logic and empirical observation in the critical rationalist paradigm), it does entail disciplined activity (see Gadamer 1960: 465; Widdershoven 1987: 116-117, 121). Below, we examine the main implications of this comprehensive activity.

2.2.3 The Double Interpretation Challenge

Whatever method is used, the aim of the interpretative approach is to obtain insight into what people in social situations do by trying to understand their intentions and motives. Social scientists and humanistic scholars attempt to access people's *common sense* thinking, to view their actions and their situation from their inner perspective (Bryman 2004: 14). They do this through rational interpretation, i.e. by finding the meaning of an action or expression from the inside, tracing the reasons for a specific action or expression. Instead of examining causes originating in the outside world, they focus on the internal reasons for actions or expressions.

Facts and phenomena studied in the social sciences and the humanities are themselves the result of interpretation. Those being studied have already interpreted the facts and phenomena under examination; their thoughts, actions, artefacts or the phenomena being studied result from their own views and analysis of the world. This interpreting of interpretations is known as the *double hermeneutic challenge* (Giddens 1976: 155, 1985: 56). Moreover, while examining the significance of thoughts, actions and artefacts, researchers are bound by the same current interpretative frameworks as are those who are being studied (ibid.; Habermas 1981a I: 177). Researchers have no exclusive access to the phenomena they study. They cannot claim to have privileged entry into a world of objective, predetermined truth; they can only use what everyone knows already. So knowledge in the social sciences is known to experts and non-experts alike. We all draw from the same, common sources for our information about these phenomena.

In this regard, all scientific interpretation depends and builds upon pre-scientific interpretation. It clarifies, once more, that knowledge is not simply objective. Knowledge circulates in an interpretative cycle, and there is no clear distinction between knowledge obtained by researchers (subject) and knowledge obtained by those being researched (object) (Giddens 1985: 56; Coenen 1987: 156, 192). The idea that subject and object occupy separate worlds is obsolete (Latour 1993; Ochs et al. 1998). There is no objective stance from which to operate, not even if the object is a phenomenon or process that took place in the past or if it concerns a person or group no longer living.

Similarly, it is futile for researchers to try to avoid being infected by the thoughts and actions of whomever they are studying. What lay people know should not by definition be considered 'polluted'; it may comprise reasonable though often limited and mainly practical insights into everyday matters. Lay people often know more than they can put into words; they have what is called *tacit knowledge* (Polanyi 1958, 1966). In their everyday practice they have acquired ideas, experiences and skills that they own and yet may find hard to articulate.

If we conclude, as we just did, that subject and object play an equally important role within the social scientific knowledge process and in humanities studies, this has far-reaching implications. These implications are not ones that every researcher is equally likely to accept, as we will see in the next section.

2.2.4   One-Sided Interpretation versus Reciprocal Adequacy

Within the domain of the social sciences and the humanities, scientific knowledge can be said to have a founding function in relation to ordinary, common sense experience. By making explicit matters that we take for granted in everyday life, scientists and scholars can clarify, articulate and enrich experiences and create (new) knowledge. Critique and reflection can thus be seen as preconditions for the acquisition of scientific knowledge. These critical and reflective activities may enable lay people to translate the pre-conscious rationality of their thoughts and actions into more conscious forms of rationality (Widdershoven 1987: 186-191). Thus the source of scientific knowledge – the self-images, insights and actions of the persons under study – can be transformed by the interpretative process in which researcher and researched engage (ibid.: 175-176).

So, while scientists and scholars definitely have something to offer, they can no longer be seen as objective know-alls in charge of knowledge creation (cf. Steiner 1991: 1-2; Guignon 1991: 100). Social scientist and humanistic scholars may have created their own vocabulary with new technical concepts and jargon (as this book demonstrates), yet the knowledge conveyed in this scientific language does not differ fundamentally from the knowledge conveyed in everyday language. Knowledge is only possible in concrete contexts where it is generated by networks of meaning, normative frameworks and power relations grounded in ordinary communities. There is no reason to believe that the language game played in science is any different from games in other social situations (Bourdieu 1975, 1989; Bourdieu & Passeron 1990).

Does this mean that a lay person's view and a scientist's view are of equal value? What is then the point of scientific research? Many scientists who lean towards the interpretative approach are reluctant to address these questions and to consider the implications. More often than not, determining the reliability or adequacy of knowledge gained within the social scientific domain or the humanities is still a one-way process. The common sense interpretations are tested against the scientific interpretations, but never the other way around (Habermas 1981a I: 193). Assuming that, ultimately, the scientists are the ones who can come up with the best interpretations of what is really, objectively going on in the minds of their research objects (the researched), the researchers take their own interpretations as the point of departure in their search for truth and rationality. So despite the diverse assumptions that distinguish interpretivism from critical rationalism, in the end they still rely on the standard concept of objectivity.

Not all researchers avoid the challenging implications of the interpretative paradigm. Some have taken the consequences for scientific research to its extreme and have developed a new approach (see e.g. Coenen 1987, 2001; Gibbons & Nowotny 1994, 2016; Bergmann et al. 2005; Regeer & Bunders 2007; Godeman 2010). Rather than viewing those they research as naive sources of information, they acknowledge that the knowledge of the researched (the knowledge object) is valuable and complementary to their own knowledge. Both the researcher and the researched can contribute significantly to the quality of the results. Instead of viewing those they research as objects, they see them as co-subjects and involve them as co-researchers in a joint research and learning process. Some researchers conclude that the knowledge acquired through this process can only be considered adequate if the researcher's interpretations are acknowledged by those being researched. This knowledge, which can be labelled as 'reciprocal adequate', is considered the most objective (Coenen 2001: 66). It is certainly a fundamental break from the standard concept of objectivity.

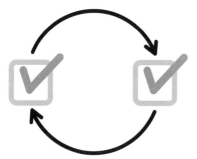

Figure 2.8   Reciprocal adequacy of research results as an alternative principle of objectivity

In the first version of the interpretative approach, the concept of valid knowledge is no different from the concept proposed in the critical rationalist paradigm. Basically, the researcher is still the objective know-all, while a disinterested search for truth is still the ultimate criterion (Kunneman 1986). The interpretative method is just one more option in the ruling paradigm – an option that may be severely criticised, since it is unclear why knowledge claims generated in this way should be regarded as anything more than subjective opinions formulated by the researcher. In the second version, the implications of abandoning the traditional 'knowledge subject'/'knowledge object' distinction are explicitly addressed, resulting in a different criterion for valid knowledge. Only this version offers an interpretative approach that can be said to form a genuine alternative to the critical rationalist paradigm.

2.3 Current Models and Future Thinking

The standard model, the dominant paradigm in modern science, evolved over more than a century. So, too, the interpretative approach, which dates back to the late nineteenth century. The twentieth century saw the birth of yet another way to obtain scientific knowledge. Since the pragmatic turn in the philosophy of science in the early 1970s and the digital turn in the 1980s, scientists have shown an increasing interest in the use of models. This is a major development that deserves special attention in our context of (complex) systems thinking.

As a relatively new aspect of science, the status of models is a matter of debate (for a review of the positions, see Winsberg 2010 and Saam 2017). While this is not the place to join the debate, it is worth looking at the uses to which scientists put models.

2.3.1 The Model Cycle

Models such as diagrams, equations, visualisations, three-dimensional images and computer simulations can be defined as idealised structures that represent actual situations. The goal is to gain insights into a complex system by understanding a simpler, hypothetical system resembling it in relevant respects (Gilbert & Troitzsch 1999: 1-25; Cilliers 2001: 138; Giere 2004: 748; Godfrey-Smith 2006: 725-730). When building a model, we use any type of knowledge available to us, gathered from all kinds of sources, whether mathematical formulas, empirical data, rough or refined theories, moving or still images, or textual descriptions. Models can help us learn more about physical systems – a climate model, for example, or a model about social systems, like a conceptual model that shows the basic food chain patterns. Models can also extend beyond their discipline, for instance a model of the food system incorporating agricultural aspects of food production as well as societal aspects of access to food and food distribution. Even in the humanities, models are used, such as mental maps of controversies relating to the current status of democracies (see Berry 2011).

Creating a simulation model begins with specifying the question that requires answering. From that comes a definition of the phenomenon or process we are interested in, the *target system*. Normally, some observations of the target are needed to provide the parameters and initial settings for the model.

Next, based on a few developed assumptions, the model is designed, usually in the form of a computer programme. Then the simulation itself is performed by executing the programme, and the output is recorded (Gilbert & Troitzsch 1999: 17-25, see figure 2.9). During this modelling process, two intermediate steps are required: a debugging to ensure that the model is correctly implemented and working as intended, and a sensitivity analysis to check how sensitive the model is to slight changes in the parameters and initial conditions.

The final step is that of *validation* of the model by testing whether its behaviour does indeed correspond to or at least show similarities with the behaviour of the target. As in conventional experiments, models can yield hypotheses, and a certain model may appear to suit the target system. Hypotheses may be proven false or confirmed by comparing the model's performance to the actual target system (Giere 2010: 273). Note that, just as with theories, in practice the verification or falsification is not absolutely separate from the validation process. Determining whether or not the output of a simulation resembles the behaviour of the target system is difficult to distinguish from whether or not the model is an adequate representation of the actual system (Winsberg 2010: 19-24).

Models are seldom merely a quantitative description of a single system with a single set of initial conditions, especially where complex systems are concerned. They usually attempt to summarise the basic qualitative features of a whole class of structurally similar phenomena, such as emergent relationships in certain aspects of a system, threshold values of important parameters, characteristic mechanisms or coherent structures, and patterns of interaction and competition between such structures (Winsberg 2010: 17).

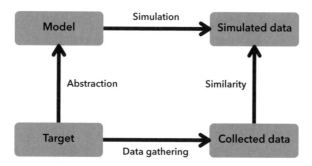

Figure 2.9 The model process
Source: Gilbert & Troitzsch 1999: 17

Complexity brings many new tools to the study of the physical and social world: sophisticated mathematical and computational methods such as system dynamics and world models, complex network analysis, microanalytical or multilevel simulation, *cellular automata*, agent-based models, and learning and evolutionary models (for an explanation of these methods, see Winsberg 2010: 26 ff).

An obvious use of models is as an instrument for building theory. Models can present a theory by illustrating the general laws and underlying principles that may apply to a particular system (Knuuttila 2011). Figures 2.2 and 2.4 are examples of explanatory models illustrating theories in biology, showing the simple, non-digital use of models. However, since models are usually designed to represent specific situations, they are never as general as a theory (Bailer-Jones 2003). And unlike formal theories, models are usually presented in non-linguistic forms without involving a system of statements (Nersessian 1992).

Models are also used as tools for exploring a theory that is already in place. For example, since we can rarely see evolution in action, evolutionary biologists use models to investigate and find proofs, and to argue for or against the plausibility of a proposed mechanism in the natural world. This indicates another function of models, namely as a tool for experimentation. In this capacity, models can be deployed both for testing existing theories and for exploring processes that these theories cannot explain (Morrison & Morgan 1999).

Models therefore play a role similar to the empirical and hermeneutical cycle: they are used for developing, exploring and applying theories and as a tool for experimentation. Models serve to satisfy epistemic purposes like description, explanation and prediction. Scientists use models to externalise and organise their knowledge of a system, enabling them to reason within a theoretical framework and evaluate their ideas (Knuuttila 2011; Gouvea & Passmore 2107). Used this way, models reflect the same inductive and deductive styles of reasoning employed in the empirical and hermeneutical cycle. Yet models may also be useful in another aspect of our effort to come to grips with reality, as the following section reveals.

2.3.2   Simulation as a Way to Enhance Systems and Design Thinking

A feature of models is their use as a tool to explore implications of theories in real situations (Morrison & Morgan 1999). When investigating if and how intervening in a system changes certain properties, models are a useful tool for examining how applying scientific concepts and principles would modify and transform actual situations. In this respect, models relate to *design thinking*: a form of solution-based thinking in which systems thinking is used to develop creative strategies to produce constructive results (Van Assen et al. 2017). Making models is a kind of thought experiment, a form of simulative model-based reasoning in which mental models of the target system are manipulated (Nersessian 1992; Brown 2004; Beisbart 2012).

Used this way, models are more about abduction and retroduction, a form of reasoning based on the principle of concomitance. The simplest and most likely explanation is sought by looking for remarkable co-occurrences or striking similarities in patterns between the model and the actual system. Like induction, this is not based on valid logical inference, but it is all that remains when only a limited set of observational data is available. Comparing the verification/falsification and validation process with benchmarking or calibration, as Winsberg (2010: 22) does, is illuminating here. Researchers can claim their proposed model is reliable by showing that the simulation matches what is known about the actual system – by comparing the results of the model to actual data, to (other) experiments, to analyses or other simulations.

An interesting example is the first macro-econometric model that Jan Tinbergen constructed to represent the Dutch economy and to devise the best solution to the Great Depression of the 1930s. The model was first sketched using empirical data and then simulated to analyse the effects of six different interventions in the economy. These simulations enabled Tinbergen to compare the actual impact of the proposed interventions in the world represented by the model. He recommended that the Netherlands withdraw from the gold standard, a policy that was indeed adopted by the Dutch government. Models such as this econometric model are designed to represent both theory (in this case, macroeconomic theory) and practice. Yet their main purpose is not to examine theory but to explore actual past and future conditions and perhaps to change these (Morrison & Morgan 1999: 23 ff; Grüne-Yanoff & Weirich 2010: 44). Simulations enable us to learn about how the world works by suggesting how or why a process develops or why a system behaves in a certain way. As the following chapters show, this feature of models can be helpful when searching for solutions to urgent problems and can generate change in society.

Simulations offer other possibilities too. While a model's internal composition represents the structure of an actual system, it can also be a springboard for envisioning potential and even physically impossible worlds (ibid.). The prospect of a new kind of science enabling the systematic study of possible worlds is relevant to the scientific understanding of ecological and social systems (Batty & Xie 1997: 175), and to our efforts to improve these systems. This is exactly what simulations are meant to do; they provide a way to visualise virtual realities and potential futures. In the context of the prospect of vision development presented in this book, this is a promising avenue that certainly deserves elaboration. The question of how simulations can play a role in the development of future *scenarios* will be saved for the last chapter, though, where the future will be our explicit focus of attention.

2.3.3 Computation and Design: New Imperialism or Emancipation of the Sciences?

We closed the introductory chapter with the question of whether the computational turn or at least the broader complexity turn can offer a new approach to tackling complex problems, or whether it is just another form of disciplinary imperialism bringing the social sciences into the realm of the natural sciences. Which option prevails depends on how computational methods, systems models and design thinking are actually employed in science. If the new computational approaches are old wine (atomistic and reductionist thinking reflected in the machine metaphor) in new bottles (the same kind of mechanical thinking in computer language with new metaphors), the former may hold true, i.e. they can be regarded as just another form of disciplinary imperialism bringing the social sciences into the realm of the natural sciences. This was certainly true in the early days with the first, simple models, when variables were still assumed to be in a linear relationship (Gilbert & Troitzsch 1999: 1). Now, with the introduction of complexity thinking, simulation models have become more sophisticated, and aspects such as non-linearity, self-organisation and emergence are taken into account – or at least that is what modellers try to do (ibid.: 10-12).

Another pitfall to be avoided is the tendency to extract external representations – whether drawings, graphs, three-dimensional structures or simulations – from the context of modelling. We should always realise that models are what they are: representations of reality but not reality itself. If we forget about the context, models stop being models and start to function as inert *reified* knowledge (Gouvea & Passmore 2017: 55-56). In this type of abstracted, extracted knowledge, there is no room for self-organisation and emergence; it has few benefits and may even be harmful.

A rather interesting difficulty with models is related to the predictive function they can perform. In the early days, simulations acquired a poor reputation because their results could depend heavily on specific quantitative assumptions about the model's parameters that were rarely supported by evidence. Apart from these teething problems and the inherent difficulty of making predictions in the social sciences (which lack the kind of general laws of the natural sciences), a more fundamental objection relating to the social sciences and the humanities is that predictions can affect results. In other words – and this is characteristic of human action – forecasts can become self-fulfilling prophecies, when people start acting according to the predicted outcome (Gilbert & Troitzsch 1999: 6). This can cause a prediction to become true exactly because the expectation was explicitly expressed. While this may be a problematic polluting factor in science, it can be an advantage for public policy. This is discussed further in the final chapter, which addresses vision development as an instrument for realising beneficial changes.

This curious characteristic of self-fulfilling prophecy alludes to the last and perhaps most important issue: the role of agency in models. Human beings have the capacity to direct their own future through action (or indeed inaction). In this respect, systems and design thinking with their methods of modelling and simulation can be useful tools, but they cannot succeed if agency is ignored. Even if it is included, as in agent-based models, this may still be insufficient. Agency expresses the purposive nature of human activity and relates to ideas such as intent, free will and the ability to strive for goals. When applied to agents in computer programmes, agency is generally much weaker in scope.

The tendency to attribute software agents with purposive intent can cause some philosophical confusion. While digital agents are often attributed with a certain intentionality in the sense that their behaviour is explained in terms of belief, desire, motive and even emotion, they are constructed to simulate simplified aspects of human intentions. Computer agents are typically imbued with properties such as autonomy, the ability to interact socially with other agents through a form of language (a computer language, not a natural language), reactivity (implying that agents are able to respond to what they perceive) and proactivity in the form of goal-directed behaviour (Wooldridge & Jennings 1995; Gilbert & Troitzsch 1999: 174-175; Grüne-Yanoff & Weirich 2010: 32-34). But digital agents do not have the same abilities as human actors.

Still, there are many scientists who think that computer simulations have valuable features, one of them being that they are well-equipped to represent dynamic aspects of change. Showing how different models can perform diverse functions and are used in a variety of ways in problem-solving may induce a more positive approach. Models can describe qualitative representations of alternative futures, both literally and figuratively (Morrison & Morgan 1999: 36). Moreover, simulations can help to develop an understanding of the relationship between the attributes and behaviour of groups of individuals (micro level) and global properties of social systems (macro level) (Gilbert & Troitzsch 1999: 13).

It is time now to investigate how these two – the macro and the micro level where respectively structure and action prevail – can be fruitfully brought together within science. This is exactly what we will do in the next chapter. But before we do that, let's first take stock of what we have learned in this chapter.

2.4 Unity in Diversity

It is clear now that the critical rationalist paradigm is not the only approach to science. Besides the familiar standard scientific method, another option is the interpretative approach. Moreover, a new paradigm has recently emerged in the form of complexity thinking, involving new methods such as modelling and simulation. The idea that a single scientific method applies in every domain and every situation is outdated. There is more than one explanatory model, and there are different approaches. This is a good thing. As systems thinker Gerald Midgley (2000: 171-216 & 254-256) states: reality is versatile and dynamic, hence our research methods need to be so too. Spatial analysts and modellers Michael Batty and Yichun Xie (1997: 190-191) comment that for complex systems, it is unlikely that any single approach which is dominantly superior for all applications can be developed. In fact, this kind of science is likely to resemble social theory where a multiplicity of disciplinary viewpoints and paradigms is the norm.

Batty and Xie think that the way elements behave in complex systems (their research field is cities) is too innovative to be subject to the iron laws of conventional natural sciences. In their contribution to *Social Science and the Study of Complex Problems*, David Harvey and Michael Reed (1996: 296-297) also argue that no single method can fully appreciate the complexity of social life (as does Cilliers 2004: 136), making methodological *pluralism* essential.

In 1975, Paul Feyerabend emphasised in *Against Method* that it would be unwise to prematurely decide that the standard scientific method is the only valid explanatory model whatever the situation (i.e. he took a stance against *monism*). Some actually consider the desire for unification dangerous; they regard it as scientific or disciplinary imperialism (see e.g. Olsson et al. 2015: 7, 9). In their view, this type of integration of knowledge is not a particularly useful approach and can even prove counterproductive when attempting interdisciplinary and integrated research.

To criticise this is the signal for the floodgates to open and to suggest that now 'anything goes' is unfair. The option to choose from many different methods actually requires more caution and responsibility, Feyerabend argues. Recognising that each method has its own limitations entails the responsibility to justify why in certain situations one is more applicable than the other and to explain what the advantages and disadvantages are. A pluralist approach contributes to the standard requirement of triangulation; it requires scientists to make several operationalisations of their concepts and to use diverse methods to measure these. And indeed, it appears indispensable for the study of complex problems.

The different approaches outlined in this chapter give an idea of the range of approaches needed to tackle the urgent questions at the interface of humanity and the planet. The art is to be able to switch between the macro and the micro level, between atomism and holism, between attention for structures and attention for the agency of social actors. The standard scientific method and computer models may be more appropriate for one case, while an interpretative approach may be more suitable for another. We need to make the best use of the choices at our disposal and always be aware of the limitations these involve.

Questions:

- Name the phases of the empirical cycle. When do scientists use deductive reasoning, and when do they use inductive reasoning?

- In what sense is theory formation regarded as a rational activity in scientific research?

- What is the induction problem? Put differently: why is it logically problematic when a theory is derived from the facts by inductive logic?

- What is verifiability? And what is falsifiability?

- How does the deductive-nomological model (D-N model) express the central ideas of critical rationalism?

- Popper developed critical rationalism to solve the problems relating to earlier forms of modern science. What kind of problems did he subsequently encounter with his critical rationalism?

- What problem is linked to the names of Duhem and Quine? And what is the lesson we learn from confirmation holism?

- What are the principal differences between the interpretative research methods and the standard scientific method?

- What is abduction or retroduction? And how do these relate to deduction and induction?

- What does the double interpretation challenge imply?

- Our interpretation of what happens in the world is caught in a hermeneutic circle. In this sense, the hermeneutic process resembles the empirical cycle. But what is the difference between the hermeneutic circle and the empirical cycle?

- What criticism has been brought forward against the interpretivist paradigm?

- How many paradigms are mentioned in the chapter? What are the principal differences in the underlying approaches?

- Why do some people state that the interpretative alternative merely represents an alternative approach rather than an alternative paradigm?

- What is the added value of the use of models and simulations in scientific research?

- What are the possible dangers of the use of models and simulations in scientific research?

- What does design thinking entail?

- What is meant by 'imperialism of the social sciences by the natural sciences'?

- What is meant by 'disciplinary imperialism'?

- What does a pluralist approach to scientific research imply?

3 Structure and Action in Science

In the previous chapter, we found that models and simulations can be defined as idealised structures that are used to represent the world. We try to gain understanding of a complex, real-world system by designing simpler, hypothetical systems that resemble it in relevant respects. Like theories, models and simulations function as explanatory models to get to grips with reality. But to what extent can we hold on to the idea that these theories and models adequately represent the world's structures? This question forms the topic of the first section in this chapter. We examine the various views on the representative function of theories and models and the difficulties they entail. The main objection is that they fail to take into account that theories and models need designers to be developed, i.e. scientists with at least a vague but most often a well-defined purpose for the explanatory model. This purpose is in turn based on a pre-established interpretative framework.

This points to a lesson that we have learned from interpretivism: reality is not a given but is to a large extent constructed by us, by human beings. We approach reality with certain theories that help us to make sense of our environment. This is certainly true of theories we develop about the social world, yet it is no less so of the models and simulations we build of the physical world. This claim that our quest for knowledge is inevitably influenced by interpretation does not sit well with the aim and claim of objective representation. In the second section, we present illustrative metaphors for both the representative and interpretative approach to knowledge acquisition – Mastermind and mapmaking – and compare these divergent views from a critical, complexity-oriented perspective.

In the subsequent section, we look for ways to overcome the impending clash between the various conceptions of science by introducing a new way of looking at reality. In addition to the structural dimension, the agency dimension is also taken into account, thus repairing the flaw in perspective that may have been present in the previous chapters. This more sophisticated view of reality offers a better foundation for studying the 'wicked' problems at the interface of human activity and planetary processes. Precisely this interface has inspired some scientists to propose a new metaphor – the coral reef – which may offer a more suitable way to visualise the knowledge process in relation to complex issues.

In the final section, the action dimension receives our full attention. We show how science can be used to go beyond its explanatory task and actually help find solutions for the issues that urgently need to be addressed. In addition to the standard research cycle, we point to an action cycle that is particularly interesting for researchers aiming to transform society and realise social change. We also show how design thinking and simulative models can support a policy cycle in which potential strategies and interventions are tested. With that, we hope to provide a picture of a full-fledged scientific approach ready to deal with the complex problems we need to solve.

3.1 Objective Structures or Subjective Perspectives?

3.1.1 Correspondence and Representation

Inherent to both the conventional, experiment-based scientific method and newer methods such as modelling and simulation is that we tend to think that the knowledge gained via these methods mirrors the outside world. We assume that our objective knowledge is an adequate reflection of reality.

Traditionally, knowledge about the world is substantiated when a theoretical statement can be shown to correspond with an actual situation. According to this *correspondence theory of truth*, the test of whether knowledge is true is whether elementary empirical scientific claims agree with objective reality. Observational terms derive their meaning from the direct connection they are supposed to have with reality. Because of this direct connection, it is assumed that these terms are certain (Koningsveld 1987: 66). Building up from this groundwork, an elaborate system of objective knowledge conforming to general laws can then be formed, or so it is thought (Carnap 1932/33: 228).

In this sense, a scientist's task is to reveal reality. Scientists make rational reconstructions of the world and formulate theories and explanations to acquire insight into reality (Koningsveld 1987: 59). The empirical foundation acts as an independent arbiter allowing us to decide whether the theory and the knowledge inferred from it are acceptable. This view of reality is referred to as *empirical realism*.

An analogy that illustrates this concept of reality and its consequential correspondence theory of truth is the board game Mastermind (Fay 1996: 209). In this game, players try to figure out the situation behind the screen, which is controlled by the Mastermind, i.e. the person who knows what the 'real' set-up is. The players try to break the code, i.e. discover a hidden sequence of coloured pins, by experimenting with the various coloured pins until they are placed in the right position. The omniscient Mastermind uses the small white and black pins to indicate whether the coloured pins are correct but in the wrong position or correct and in the right position (compare with falsification and verification in scientific experiments). In the end, the screen is pulled away (compare with the dis-cover-y of reality). The player gets to see the actual situation and is thus able to check directly whether the order and colours of the pins agree with the situation on the other side. The player has succeeded when there is a one-to-one match between the chosen set-up and the original order.

Figure 3.1 Mastermind

The process that Mastermind players go through shows striking similarities with the view on truth that Popper tried to capture in his criterion of *verisimilitude*. This states that a relative approximation of the truth can be found by weighing a theory's truth content against its falsehood content, taking into account the degree to which the theory has been confirmed. The latter reflects the number and severity of unsuccessful attempts to show that the theory is false (Popper 1963, 1972). According to Popper's theory, it is possible to find the best theory or set of theories that approximates the actual situation closest by using the criterion of verisimilitude. Yet if observations are always influenced by theory, and if it is never possible to verify a theory or prove that it is absolutely false, this criterion is of little value. Although philosophers of science have tried to refine and improve Popper's criterion of verisimilitude, none have been able to find a solid, undisputed criterion for measuring the truth.

The advent of new methods has revived the debate on how to characterise the way that knowledge about phenomena relates to the phenomena themselves. Today the question is: how do we manage to gain knowledge about phenomena through models and simulations? What is the nature of the relation between phenomena in the world and phenomena in a model? Some compare this to the earlier focus on describing the structure of scientific theories, and they see it as a continuation (cf. Bailer-Jones 2003; Knuuttila 2011; Gouvea & Passmore 2017: 52). But since models are usually non-linguistic rather than a system of statements, this parallel is not especially useful. Moreover, the same objections apply here as are raised against the conventional view, namely that observations are influenced by theory and that empirical evidence cannot be the ultimate touchstone of knowledge.

A more useful way to view the way models and simulations function might be to see them as an attempt to represent complex, real-world systems in simpler, hypothetical systems that match them in relevant respects. The representation may take the form of similarity, which means that there is supposed to be a generalised resemblance between the model and the real world. Another, stricter form of representation is *isomorphism*, which demands that elements and connections in the model relate one-on-one with elements and their respective connections in the real world. Here the representational relationship is viewed as the best possible match between a model and the actual situation (cf. Giere 1999, 2004; Suárez 2003, 2004, 2010; Godfrey-Smith 2006). David Byrne (2014: 160) argues that the objective of a simulation or model should be that the mechanisms demonstrated represent actual generative mechanisms. He is thus a typical advocate of the latter form of representation. In his view, the description of a system's characteristics constitutes an overarching law, and if the model's analogies of the real world hold true, it is because the generative mechanisms are of the same general form.

However, attempts to define the relationship between models and real situations in terms of structural mapping, whether in similar or (partially) isomorphic models, have not been successful. Philosophers have argued that to propose a single description of the relationship is problematic, since in scientific practice, models only partially match the actual situation, and different models do this in different ways (Morrison & Morgan 1999; Suárez 2003, 2010; Godfrey-Smith 2006; Pieters 2010).

What happens in this type of account is that representation is claimed to be solely grounded on the structural properties of the model and its target system.

> 'It is as if the things taken as representations did their representing job, i.e. created the relation of representation by themselves, by virtue of what they (and their target systems) are. This actually appears to be often the case – when the application or the interpretation of the model in question has become routine.' (Knuuttila 2005: 1269)

It is the same pitfall described in the previous chapter, namely the danger of reifying models when taken out of context so that the representation in a model is no longer treated as an actively constructed artefact through which knowledge about the real world is conveyed and models start to live a life of their own. But it should never be forgotten that

> 'it is not the model that is doing the representing; it is the scientist using the model who is doing the representing.' (Giere 2004: 747)

The real world is not neatly arranged and ready to be packaged in a model; it is the researcher's task to carve out a well-defined and clearly demarcated phenomenon. This applies to both the natural world and the social world. A model requires a purpose, and a model's structure is inextricably linked to its purpose. For it to serve this purpose, it requires a pre-established interpretative framework (Pieters 2010), because we have no stable foundation when we reduce a complex system to a model (Cilliers 2001; Giere 2010; Gouvea & Passmore 2017).

3.1.2 Perspectivism and Fallibilism

A basic counterpoint of the realistic vision of the real world is *perspectivism*. Perspectivists claim that our view is defined by our specific frames of reference. These include conceptual blueprints or theoretical frameworks that direct our focus in a particular direction and entail various assumptions. These assumptions may introduce a certain bias, so that theories can probably only ever be partially correct, under certain conditions or with limited application. The theory chosen is dictated by a person's perspective. No one can set aside their perspective and objectively assess all the available theories. Even a meta-perspective has a particular perspective at its core, albeit at a higher level. From a perspectivist view, all scientific theories are ultimately arbitrary constructions that reflect forms of our own specific perspective rather than some external objective reality.

Which Theory or Solution is the Best?
It Depends on How You Look at It!

Agronomists may frame the issue of malnourishment among children in so-called developing countries in terms of food production. Nutritionists may see it as a question of selecting an optimal diet and changing consumption patterns. Epidemiologists will view it in terms of remedies for diseases that increase the demand for nutrients or prevent their absorption, demographers in terms of controlling the rate of population growth that may otherwise outstrip agricultural activity, spatial planners in terms of more possibilities for food storage and adequate distribution, political scientists in terms of equal access to land or other food resources, and economists in terms of guaranteeing sufficient purchasing power or an equitable distribution of wealth.

In the natural sciences, the food issue is usually approached as a matter to be solved with models and calculations of production factors such as soil, temperature, water, plants, animals, labour, chemical pesticides and fertilisers. In the social sciences and the humanities, it is the demand for food and cultural preferences, emotional values or inhibitions, taboos or restrictions rather than supply that is thought to determine the general focus of the food system (cf. Fresco 2012: 37).

In short, perspectives determine how people analyse and theorise about a problem. And to a large extent, how a problem is defined determines what solution is deemed necessary or adequate. When the frameworks diverge, differences may not be easily resolved, if at all, by appealing to data. When different frameworks apply, different facts are cited and interpreted in different ways (Schön 1987: 4-5). Moreover, different perspectives and different ways of structuring an issue also lead to different views about potential research approaches.

The idea that science is biased does not mean that today's science is functioning poorly. It has nothing to do with a failure that could be corrected if only we had better researchers, better instruments, better tests or better hypotheses. It is inherent to the human capacity to know and to reason. No matter how many persuasive empirical confirmations of a favourite theory or manifest refutations of an alternative we may have, it can never be enough to guarantee that we possess the absolute truth.

The recognition that what we know is at most highly probable and can never be considered irrefutably true is known as *fallibilism*. This is a philosophical position that maintains that we can never know anything with absolute certainty, even if we have good reason to consider our knowledge to be true. This position encourages us to be modest about what we know and to avoid arrogantly assuming that we know everything (Fay 1996: 208).

Many of the most celebrated theories of the past – such as Ptolemaic astronomy, mercantilist economics, and humours as an explanation of disease – have been shown to be so pertinently untrue that no one gives them credence any longer. Yet that is no reason to be disdainful, because a century from now people will doubtless regard many of today's finest scientific achievements with amusement. They may wonder with a smile whether we really thought that germs caused disease or that the money supply determined price levels or that we believed in the subconscious, to name just a few examples. Moreover, it is entirely possible for discredited theories to experience a revival and rise again to prominence. After all, they may lead to revised theories with far more potential than any of the other available theories (ibid.).

A strategy that employs a masked form of fallibilism centres on making less ambitious claims. Instead of proclaiming universal laws in *deterministic* claims, this approach is about proposing more modest statistical claims, showing empirically observed regularities based on probability. This is common practice in science today. Mendel's laws are an example of a statistical claim, although the label suggests otherwise. And this is exactly where the danger lies: statistical regularities are too often presented as universal laws, just as correlations are presented as causal connections while such a strict relation doesn't actually exist.

3.1.3 An Instrumental Outlook on Science

Today, few scientists hold on to the empirical realist view that knowledge claims can be assumed to correspond with the actual structure of the real world. Most scientists agree that the *objectivist* approach – in which phenomena are treated as external facts unaffected by human influence – is untenable. Acknowledging that the phenomena we study can never be understood entirely free of our own perspective, they accept that it is impossible to produce an absolutely true description of reality. They take a more modest attitude and see theories mainly as tools for making predictions.

Similarly, scientists do not always use models as representations of the real world. Rather than reducing them to representations of some pre-existing natural or social systems, it may be more realistic and useful to see models as epistemic tools to help us make sense of the world. Models are designed in such a way that they enable the investigation of certain systems; we can learn more about the system by running simulations and by playing with the parameters of the model. Models provide supportive tools with which to draw inferences about the world; they enable a form of simulative reasoning by playing with mental simulations of the target system (Nersessian 1992; Suárez 2004; Knuuttila 2005, 2011; Beisbart 2012). They also inspire new ways of looking at phenomena and encourage new kinds of questions.

Since the knowledge generated by scientific theories or models of reality and the instruments used to measure and analyse that knowledgeare treated as if they were the same, proponents of this view are known as *instrumentalists*. Grand-maître Popper is said to have tended towards the moderate instrumentalist conception of knowledge, although in his criterion of verisimilitude he did not reject the underlying ideas of the correspondence theory of truth entirely. The tendency to cling to realistic assumptions, albeit often implicitly, also seems to apply to a new generation of philosophers of science who remain loyal to realism and yet lean towards a more pragmatic, instrumentalist perspective (e.g. Morrison & Morgan 1999, and Byrne 2014: 160). While they are theoretically committed to an agnostic view of the true nature of reality ('we do not know'), a certain outlook on the world still pervades from their implicit assumptions. This is probably a product of the basic, supposedly obvious assumptions that the only way to obtain scientific knowledge is by building on empirical observation and logical inference. So reality is thought to be a 'given' after all; it is supposed to have a structure that exists independently of human knowledge. This certainly resembles a realistic vision of reality.

However, like empirical realism, the instrumentalist position offers no escape from the objection that the empirical basis fails to provide the desired certainty. If all observational claims depend on theory, it helps little to shift the criterion from 'corresponding with the facts' or 'representing reality' to 'predictions coming true'. Unfortunately, the instrumentalist approach to science is not a solution either, since it shows no way out of the problems we have encountered.

Yet from another perspective, instrumentalism does offer a solution. This becomes evident when it is viewed as an approach focusing on the pragmatic and action-oriented aspect of science. In an instrumentalist perspective, the focus of the knowledge process shifts from uncovering the truth through observation to realising theories (Dutch: from *waar-heid* via *waar-neming* to *waar-maken*; German: from *Wahr-heit* via *Wahr-nemung* to *wahr-machen*; French: from *la verité* via *verifier* to *realiser*). Rather than finding solid-based knowledge – the 'know-that' that is the substance of accepted knowledge claims – instrumentalists are interested in gaining 'know-how', i.e. knowledge that enables us to work things out. For even if we could sketch a 'realistic picture', it would frequently not help much since systems change unpredictably over time – often due to our interaction with it (Törnberg 2017: 24).

But perhaps it's better not to put the positions in opposition to each other; perhaps we should be viewing them from an 'and-and' perspective instead of an 'either/or' perspective. Margaret Morrison and Mary Morgan (1999: 30) argue that rather than being in conflict, the instrumental and representative functions of simulations are complementary. While simulations are a way to apply models, they can also serve as a bridge between abstract models (with theoretical facts) and technical contexts (with concrete facts). Thus, models are able to represent physical or social systems at two distinct levels. The first is the level of concrete detail through the kind of simulations that models enable us to produce; the other is the higher level structure that the model itself embodies in an abstract and idealised way.

A good model is one that fulfils the purpose for which it was designed. For instance, a scientist investigating diffusion might require a model showing water as a collection of molecules, while a scientist examining how water flows through pipes might design a model in which water is shown as a continuous fluid. While a simple schematic model of Milankovitch's theory may be helpful to gain insight into climate change (see figure 2.6), a more useful model to learn about the potential effects of climate policies would be a computer simulation programme. Models of the same thing can thus vary depending on their purpose. The type of model used to represent a system depends on the type of problem we need to tackle (Morrison & Morgan 1999).

The most that experimental methods can reveal is how well a model fits aspects of the system it represents. This can be tested by establishing whether a model 'works' or not. That can be difficult; it may not be easy to reach a generally agreed assessment. The measure of adequacy may depend on individual or communal understandings of current standards in particular disciplines or on the model's purpose. The ultimate criterion for judging a model's success is its effectiveness rather than its correspondence with the target system (Giere 2010; Olaya 2014; Gouvea & Passmore 2017). In this sense, an instrumental view of reality can help us to move away from a rather problematic concept of truth. By shifting the focus of the knowledge process from 'knowing that' to 'knowing how', the instrumentalist perspective offers a way out of the dead-end street of empirical realism. In the final section of this chapter, we will return to this issue and further elaborate on the pragmatic and action-oriented aspects of science.

3.2 A Clash of Approaches?

Evidently, the critical rationalist and interpretivist paradigms represent different views of reality and how to obtain knowledge about that reality (for a summary of the positions, see table 3.1 in section 3.2.3). It is no exaggeration to say they seem to entail diametrically opposed ontological and epistemological assumptions. In this section, we examine the implications of these stances by analysing the metaphors that can be used to illustrate the positions. This leads to the conclusion that both the representative and interpretative view of knowledge acquisition have their pros and cons. The major question is how best to combine these to get the most out of the advantages while avoiding the pitfalls that lurk in both.

3.2.1 There Is No Mastermind

Knowing what we now know, we can only conclude that empirical realism and its concurrent objectivist knowledge theory are problematic. The correspondence theory of truth or any kind of representational theory of knowledge paints too simple a picture of the relationship between the person who knows (the knower) and what that person knows (the known). Empirical realism can therefore be dismissed as naive realism. The Mastermind analogy does not hold for a number of reasons, of which the most important are as follows.

Brian Fay (1996: 209) points out that in Mastermind, the players know that the basic materials are pins, colours and holes. These are the building blocks. However, in science, the structure is not a given – it is part of what the scientist needs to find out. Unlike in Mastermind, scientists have to discover the basic materials that constitute the terrain.

Moreover, scientists are unsure about how the terrain is organised. In Mastermind, players know that the pins must be arranged in a certain order. But scientists have to decide what order is, and in what way the terrain is ordered: is order a mathematical formula, a law or evolutionary explanation, or a completely different type of order? That is why observations are always influenced by theory: scientists have to use theoretical concepts to guide their research if they are to obtain empirical facts.

Even when the basic material and order of the terrain are known, another crucial difference remains. Unlike Mastermind players, scientists can never remove the screen to find out what reality really looks like. We can never step outside of our own mind to test whether our statements correspond one-on-one with reality or whether our mental model of reality (the sequence of the pins on the board) corresponds with the real world (the sequence of the pins behind the screen). No matter how much we try to test the relation between an empirical fact and our truth claim or between our model and the real world, there can never be a direct connection. An unmediated, direct access to reality is impossible. The facts themselves are unattainable.

In fact, we can never be sure whether it is even possible to detect some sort of master code, some sort of order. After all, we can never know if there really is a reality 'out there' waiting to be 'dis-covered'. In Mastermind, the code breakers know there is a master code; this is part of the game and all the players know the rules. In real life, scientists can only assume that there must be some kind of order. And this hypothesis may be incorrect.

Then there is the matter of defining the system and setting boundaries. For many phenomena, it is difficult to give precise descriptions and to draw clear boundaries around them. It may prove impossible to find generally accepted definitions of a red or yellow pin, or even of a pin. How we determine the boundaries of an ecosystem or how we define 'ecocide', for example, is a matter of debate. Definitions of a phenomenon and the boundaries of a system may vary with the perspective of a given discipline, and they rarely carry universal agreement. In the course of a scientific inquiry, the boundaries may be expanded or narrowed down (Gouvea & Passmore 2017: 54-55), leading to adjustments in the description. This is certainly true of complex adaptive systems, which continually change due to constant interactivity both within the system and with the external environment. It is precisely this interaction with the surrounding environment that obscures the boundaries, making it difficult to demarcate the phenomenon (Törnberg 2017: 29). Whatever description and boundaries we choose determines the frame of the system.

All this leaves us little option but to discard the Mastermind analogy. Since we can never establish an indisputable version of reality, we can never be certain that our knowledge claims and simulative models actually correspond to reality or represent it adequately. Since there is no ultimate criterion to determine the *validity* of knowledge claims and simulative models, we must therefore take fallibilism seriously. How, then, do we describe the scientific knowledge process adequately? To answer that question, we invite the reader to consider another analogy.

3.2.2 From Mastermind to Mapmaking

Instead of viewing the scientific knowledge process as an analogy of Mastermind, Fay (1996: 209-210) suggests we view it as cartography, as a matter of mapmaking. Here, we investigate what this mapmaking analogy entails.

First, it should be noted that various aspects of the terrain that the cartographer plans to map are not simply 'given'. It is often the cartographer's own interests and the purpose of the map that dictate this aspect of the process. Maps of the same area may highlight topography or population, roads, vegetation, the distribution of wealth or any other characteristic that scientists find of interest.

Second, the type of representations used is determined by certain conventions (what should be represented and for what purpose?; where and at what level do mapmakers view the terrain?). Maps of the planet made from the moon's perspective differ significantly from maps made on Earth; two-dimensional maps differ from three-dimensional maps, maps made using Remote Sensing differ from maps made with Gapminder; and so forth. Thus, a mapmaker's tasks are significantly different from those of a Mastermind player, for whom the form of the representation is determined by the rules of the game.

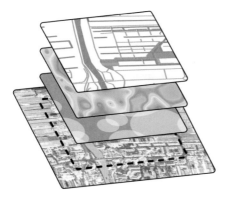

Figure 3.2   Different maps of the same area

Third, unlike the clearly demarcated terrain of Mastermind, within cartography there are no clear boundaries. There is no blueprint to check whether a map is correct. In Mastermind, the constituents (the basic material in the form of pins) and the way they relate are assumed to exist independent of the players. In cartography, the terrain is an entity that is at least in part defined by the chosen form of representation. For instance, a cartographer may decide to make a map of eco-systems and not include socio-technical systems (perhaps because these are not considered 'natural' and are therefore not eco-systems). Whatever falls outside the cartographer's concern or field of interest will not appear on the map.

Unlike in Mastermind, in cartography it is not assumed that the terrain has already been charted and the map is waiting to be 'dis-covered'. Mapmakers are not searching for the One True Map; the idea of a pre-given world is meaningless and plays no role. In cartography, there is no best map of a given area, because an endless variety of usable maps can be made of any area. How the map is made depends on which aspects of the area are represented and the way these are represented, all depending on the map's purpose and the perspective of the mapmaker.

While in Mastermind, a player tries to 'dis-cover' the inherent, pre-given structure that is represented by a row of pins arranged behind a screen, a mapmaker's task is to translate the world we learn about through our senses into some form we can understand. In Mastermind, the question is whether the hypothesis (the player's attempt to work out the sequence) agrees with the actual situation (the order behind the screen); in cartography, conceptual assumptions determine what the order is and hypotheses are judged according to their rational acceptability. All this is governed by a continually developing scientific knowledge process and developments in society at large. In the game of Mastermind, comprehensibility is a quality attached to a particular matter that exists independently; in mapmaking, comprehensibility involves comparing different hypotheses in terms of their capacity to provide insight into particular phenomena and to demonstrate ways of dealing with certain problems.

In short, maps of the world help us to gain insight into reality, to develop explanations and to discover coherence in these explanations. Different types of maps are possible. So maps are never complete; they can never represent the real world in its totality. But they do provide useful insights. Maps do not convey knowledge that exists independent of us; it is knowledge that is closely connected to the choices we make in view of our aims and preferences. We ourselves determine what maps and what particular approaches or methods we prefer to use to explain the world.

3.2.3 Coherence and the Explanatory Power of Narratives

This thought experiment on mapmaking demonstrates that it is up to scientific researchers themselves to create order in the reality they study and to establish coherence in their data. Justifying knowledge claims is less about the relation between individual statements and reality and more about the coherence between applicable statements in a system of wider importance. It is not about a perfect fit between a model and its target system but about the mutual coherence of several models of subsystems that together try to capture a larger system.

Unlike correspondence or representational theory, this approach employs coherence as a criterion for assessing the validity of knowledge claims or models. The *coherence theory of truth* is about assessing whether different statements or models of the real world combine to form a consistent whole and produce a viable explanatory theory. The result needs to be a coherent narrative demonstrating how the real world works. In this view, the truth of a claim or model depends on the extent to which it agrees with a larger system of claims or models. Meanwhile, its truth is also measured against current views about the accuracy of the used methods or instruments.

Coherence theory is not without its problems either, unfortunately. A major issue is that the internal cohesion of a set of statements or models is not sufficient to show that they are true. That a narrative appears to be consistent is not a guarantee that it correctly represents the real world. We may try to assemble the best possible explanatory model of the real world by modelling and experimenting, theorising and adjusting maps as best we can, but whether this brings us any closer to the truth is impossible to know with certainty (cf. Suárez 2003, 2004, 2010).

Another difficulty is that if narratives and explanatory models contradict one another, how do we determine which is correct? It is entirely possible for two different theories or models to be internally cohesive yet mutually incompatible (see Bailer-Jones 2003). One of these has to be false. Whether we can compare explanatory models, and if so how, is a complicated question. It requires further consideration and is dealt with in detail in chapter 5. For the present, let it be clear that coherence theory is accompanied by a fallibilist knowledge theory.

These difficulties may lead us into a position of extreme *relativism*. The ostensibly stable rock-bottom foundation of critical rationalism turns out to be relative, open to nuance and ultimately based on fallible assumptions. The narratives that the interpretivist approach produce may form superbly aligned structures yet may be 'mis-fits' or may be entirely wrong. And in the worst case, the maps we thought would form useful representations of the real world may simply be erroneous.

While this pessimistic view of the scientific knowledge process is neither inspiring nor hopeful, there are, happily, ways to escape the situation. Popper offers the first encouragement to avoid this pessimistic trap (in his first publication in 1934: 66-67):

> 'Science does not rest upon solid bedrock. The bold structure of its theories rises, as it were, above a swamp. It is like a building erected on piles. The piles are driven down from above into the swamp, but not down to any natural or 'given' base; and if we stop driving the piles deeper, it is not because we have reached firm ground. We simply stop when we are satisfied that the piles are firm enough to carry the structure, at least for the time being.'

If we follow this strategy, we may be able to avoid both the Mastermind illusion and the danger of sinking into quicksand.

We can conclude from the above that the boundaries of a phenomenon or a system are generally unclear and that we frame these by describing them from a certain perspective. Yet we cannot build or frame whatever we like: reality constrains where the contours can be drawn (Cilliers 1998, 2001). The properties of theoretical statements and scientific models may exist independent of the scientists who use them. These 'objective' properties are evident in the concrete effects of the statements and models – sometimes even in effects of which the originators are not aware and which were never intended. So the boundary is neither merely a construction nor a purely natural, objective thing – it is, rather, a mix and an ongoing interaction between these (Richardson & Lissack 2001; Törnberg 2017).

For many scientists, tangible external effects provide sufficient evidence to support the idea that scientific theories have an objective structure outside the minds of those who developed them. They are not ready to give up the idea that statements may have objective attributes independent of specific personal assumptions and perspectives held by individual scientists. Similarly, they prefer to maintain that the network of truth claims represented by a body of knowledge in a certain phase of scientific development reflects these objective properties (Chalmers 1999). As André Klukhuhn (2008: 241 & 310) encapsulates in a powerful image:

'While facts may be negotiated knowledge and not part of the real world, they are our steady points of reference in our knowledge of the world, extracted by the world: the pegs with which we hang our theoretical networks to wave in the cultural wind on the washing line of the world.'

Klukhuhn believes we should accept that the factual pegs are fastened less tightly to the washing line in the social sciences and the humanities than in the natural sciences (ibid.: 313), a position many scientists will agree with.

3.2.4 Towards a Network Model of Correspondence and Coherence

What we have learned from the above is that a fallibilist approach is not incompatible with the concept of an independently existing world structure. Fallibilism does not deny the existence of an independent world 'out there'. It merely states that, since our capacity to know is limited, we can never know whether we have actually managed to capture this structure with our scientific theories and explanations.

This combination of perspectivism with a fallibilist approach may offer an opportunity to proceed beyond the traditional opposition in which the natural sciences are viewed as realistic and objectivist while the social sciences and the humanities are associated with constructivism and subjective perspectives. Such a division often produces an oversimplified view that one sort of science can lead to genuine knowledge while the other only leads to knowledge claims of an extremely relative nature where one perspective is as legitimate as any other.

On closer examination, objectivism and relativism are not necessarily contradictory (Latour 1993: 111-112, 1999: 73; Fay 1996: 220). They actually operate in the same spectrum, although at opposite ends. Both assume that objectivity requires a direct connection to the real world – the only difference being that from the objectivist perspective, it is possible to connect immediately (i.e. without mediation by any guiding perspective), at least theoretically, while the relativist position denies this possibility. Objectivism and relativism are therefore two sides of the same coin. We can escape the dilemma of having to choose between objectivism and relativism by simply refusing to see them as two mutually exclusive alternatives.

So the way to escape from this opposition is not to link perspectivism by definition to extreme relativism but instead to connect it to a fallibilist perspective (ibid.; Latour 1993: 113, 2017: 24; Bernstein 1983: 69). Combining perspectivism and fallibilism offers an approach to knowledge that does not depend on the principle that science must be able to present a one-hundred-percent valid picture of the one true reality. While in this view of the knowledge process, the role of perspectivist elements is recognised, it does not extend so far as to suggest that one theory is as good as any other or that it is impossible to assess the quality of a theory since different theories cannot in essence be compared.

Adopting the mapmaking metaphor instead of the Mastermind metaphor does not mean that one map is as good as any other, nor does it mean that it is no longer possible to differentiate between good and bad maps. So we needn't be afraid that abandoning an objectivist and realist view will lead to the gradual demise of the scientific knowledge process. Adopting a fallibilist approach as a realistic one need not result in such a doom-laden scenario or in absolute relativism or even nihilism.

We can certainly argue that accurate and inaccurate maps and models exist, or indeed maps that clarify and maps that confuse, or useful and useless maps. Mapmaking is not an activity we can approach without due preparation. It is delineated by fact (i.e. data that we assume to be true since they have not been shown to be false) and by practical requirements for their construction and use. Maps can be relatively good or bad, more or less reliable, detailed, inclusive and useful, and more or less explanatory without having to be the One True Map, the blueprint by which all other maps are measured. Maps can be constructed and evaluated well enough without recourse to an ultimate blueprint.

Therefore, it may come as no surprise that, in practice, the scientific research process employs a network model (see Latour 1993: 117) in which correspondence and coherence theories of truth are not viewed as contradictory but rather as complementary. It borrows different elements of both, while setting other elements aside. From correspondence theory, it incorporates our experience that facts are established aspects of our knowledge about the world which are wrested by the world itself and which we cannot bend to our will by our own mental effort. At the same time, we have had to accept that our knowledge of the world is always mediated and negotiated: we form our own image of the world and express our ideas in knowledge claims and models, but these can never represent the real world directly. The immutability of facts makes them neither value-free nor immediate. From coherence theory, we take the notion that scientific knowledge about the world cannot simply be viewed as a collection of unconnected facts; it is a network of theories in which these facts are combined in such a way as to give them meaning and structure – in particular cultural, social and paradigmatic contexts.

While in this combined perspective we acknowledge that scientific descriptions or models of reality are social constructs, we also acknowledge that these constructs are made and shaped by the real world (Byrne & Callaghan 2013: 33). Different framings of the same system may be possible, but the real world does influence their structure. To show how correspondence and coherence theory combine, let us return to the image of the pegs with which we hang our theoretical networks to wave on the washing line of the world:

> 'The actual pegs are specific points on the washing line; the networks of theories held by the pegs may have many forms, colours and patterns and may billow in the cultural wind in all kinds of ways depending on how it blows, because it blows whether or not there is washing on the line.' (Klukhuhn 2008: 241)

So we can conclude that maintaining the criteria of correspondence theory and coherence theory simultaneously need not lead to conflict. They may complement each other usefully. It may be beneficial for practical research – certainly multi-disciplinary, interdisciplinary and transdisciplinary research – to create a model that examines both correspondence (aspects of our knowledge established by the real world – the pegs on the washing line) and coherence (the overall agreement of our explanations).

3.3 Beyond the Oppositions

3.3.1 The Duality of Structure

Contemporary philosopher of science Roy Bhaskar (1975, 1986, 1989, 1991) argues that it is possible to avoid the two conflicting approaches and to build a bridge between them. He offers a new, integrated vision of the relation between human agency and structures in the real world; a view that seems partly inspired by the ideas of sociologist Anthony Giddens. Both view individuals and society as interdependent, in the sense that people are influenced by the society they live in, while their actions also influence that society. Structures and actions are not separate: they are mutually interactive. The interplay between action and structure and the emergence of action from structure and vice versa is a continuous process. As a result, it is difficult to divide actions and structures, or people and society. In real life, they are inseparable.

Giddens (1985) refers to this process as the *duality of structure*. With this concept, he tries to transcend the apparent conflict between the objectivist vision and the constructivist vision of reality. Giddens argues that both are true in part. On the one hand, societal structures are embedded and institutionalised to the extent that they are almost reified and function as inescapable natural laws. On the other hand, social actors actively give meaning and form to their personal world and always have the option to act differently. When they do, the social structures and the overall system may change. Social structures therefore cause and are caused by action. They are reproduced if groups keep repeating them, and they can change if enough people start acting differently.

Take language, for example. We would find it impossible to make ourselves understood without language. A language has rules (structure) that enable us to communicate with each other. Language also allows us to express ourselves and understand the world around us without being conscious of those rules. Language exists because we constantly use it and continually reproduce the rules. Moreover, language is an organically developing phenomenon in which new words emerge and old words disappear while new phrases and rules develop. So it's both a robust ancient communication system and a continually emerging phenomenon.

Similarly, the emergence of natural (complex) systems depends both on the complementary features of structured order and on chance and arbitrary phenomena that allow space to manoeuvre (Ulanowicz 2009). The features that bring about order remind us of structures in a social system that can also behave in a law-like fashion. Between the threads of this seemingly tight fabric of causality (structures and tendencies), however, there is the possibility of escaping rigid determinism. Like social actors, ecological organisms can organise power to influence their environment. This is where agency comes in.

Thus, this perspective takes both agency and structure into account, with reality as a dialectic between them. Giddens tends to view these as two parts of the same process. The foundation of our knowledge claims lies neither simply in their reference to or their representation of objective reality, nor solely in people's minds. We should not view knowledge as existing in a separate realm but rather as embedded in our (communicative) practices, activities and conventions. Yet there are also theorists who claim that structures and agents or actors possess distinct properties and powers in their own right, and that they are a different type of entity. To ensure that one is not reduced to the other, we must keep structure and agency apart in our analyses and study the links and interaction between them (for a discussion of this position, see Törnberg 2017: 50). In the following section, we do precisely that.

3.3.2 The Stratification of Reality: From Naive to Critical Realism

Bhaskar's *critical realism* (1975, 1986, 1989, 1991) takes account of both the stable, apparently constant nature of the real world and its at least partially constructed changeable character. Critical realists understand that the categories they use to explain reality (e.g. the categorisation into physical, chemical, biological and social processes, and the invention of terms such as atom, string and DNA) are probably only provisional. Unlike empirical realists, they recognise that there is a difference between reality itself and the terms we use to describe and explain it. And they accept that these terms will therefore never be completely accurate. By extension, critical realists are also more flexible in using theoretical terms in explanatory models that are not immediately empirically observable.

Properties of Crops: Sometimes Visible, Sometimes Not

Mendel's laws, which predict how heredity works, represent a theory in which empirically observed regular occurrences are seen as manifestations of processes at a deeper and in his day still undetected level. In other words, observed regularities were explained by learning to see them as manifestations of these profounder processes. While these processes were imperceptible in Mendel's day, with modern DNA technology it is possible to see what is taking place.

Even though the theory is known as Mendel's laws, this is not a theory that applies generally, since the explanation is only statistical. So it can offer no guarantee that phenomena predicted in the theoretical explanatory model will actually be observed in practice.

Bhaskar views nature as a multi-layered, stratified structure of physical, chemical and biological processes and human action. The relation between the layers is one-directional, so that while we can say that the human body is made up of chemical substances, we cannot say that all chemical substances have human traits.

The world of human experience is more complex than the physical world. It occupies a higher position in the hierarchy than the physical world. The higher you go in the hierarchy, the harder it becomes to isolate phenomena. Studying more complex issues therefore requires a different approach than the study of issues at less complex levels. The methods that work at lower levels are simply insufficient to explain what happens at higher levels. At the same time, they are not unconnected. Part of what determines what people are and explains their behaviour is their physical biological condition. Yet the way people act can transcend this and is not simply explained by physical, chemical and biological processes of the body. And while humans are necessarily subject to the laws of physics, chemistry and biology, they are also capable of doing far more than merely obey these forces. People have the capacity to a large extent to give their life shape and direction. A different approach is therefore required to study human action and social processes.

All this becomes even more complex when we examine the interface of the physical world and human activity, for here it is even harder to determine how physical structures and human activity interrelate and what kind of interchange takes place. We may assume, for example, that the planetary system and its subsystems – the biosphere, hydrosphere, geosphere and atmosphere – all operate and would have done so whether or not human life ever developed. But now that we humans are here and now that we have taken possession of the planet, we appear to be having a significant impact on it. Science faces a huge challenge to understand the impact of human activity on the complex interactivity between nature's subsystems.

To analyse this properly, it is useful to divide the physical world into three levels: micro, meso and macro (Kunneman 2005; cf. Newell 2011). At the macro level, the physical world is the universe, which is so big and all-inclusive that the human impact on it is negligible. At the micro and meso level, human impact is certainly evident: our impact on the environment (excessive carbon dioxide in the air or plastic waste in the sea) may well lead to disastrous distortions of the subtle balance in our planetary system.

On the one hand, we are part of a gigantic overarching system that we are hardly able to influence at all, a domain in which life is determined by an *event causality* – causal impacts linked to certain conditions of which we as humans often have no knowledge and which we cannot control despite their effect on our life and our activity. On the other hand, we have the ability to influence systems closer to our own environment, whether positively or negatively. At this level, we encounter *actor causality*: in this case, humans occupy a relatively autonomous position compared to most, if not all, environmental factors. Each person is an autonomous cause at the core of his/her own activity and has the capacity, at least in principle, to decide whether s/he will allow environmental factors to influence his/her individual actions.

Here, again, we see a mutual interdependence between the human actor and the physical and social systems. This is the duality of structure in operation. And the perspective from which you perceive an issue can have a huge effect on your vision. Looking at the duality from the structural side will give a completely different outlook than looking at it from the actor's point of view. Some scientists argue that there is no need to worry about global warming because when measured in millions of years, the planet's climate is barely affected by human activity. Their perspective is that of event causality. Yet the vast majority of climate experts agree on the importance of giving serious consideration to actor causality. They point out that humans have had such a fundamental impact on the precarious balance of the planet's overall system, in an age that some have already declared the Anthropocene age, that it may eventually make human habitation impossible.

Critical realism is the missing link that bridges the gap between naive-realist and extreme constructivist visions of reality. It offers a view of reality that fits neatly with the knowledge theory combination of perspectivism and fallibilism, proposed here as a way to reconcile the traditional conflict between absolute objectivism and radical relativism or subjectivism (see Table 3.1). This introduces the possibility of a meta-position that could be a worthy successor of the critical rationalist and interpretative paradigms – a paradigm that acknowledges the need for complexity thinking to deal with the 'wicked' problems we face. This may further be enhanced by integrating an instrumental outlook on reality that abandons the problematic objectivist concept of truth and the representational function that is inherently connected to naïve realism, as we hope to be able to show in the next section.

3.3.3 A New Perspective: Knowledge as a Coral Reef

Some theorists agree that critical realism offers a subtler, more nuanced view of reality, yet they find the representative associations that are still raised by this approach problematic (see Cilliers 2000, 2002; Thompson Klein 2004; Osberg et al. 2008; McMurtry & Dellner 2014). They argue that it is misleading to suggest that complex models represent systems as they actually are or that they cover every aspect of these systems. This would be to return to the traditional assumption of a real world 'out there' whose rough outlines we know. In practice, natural complex systems have no clearly defined boundaries. At most, our models are functional or pragmatic instruments, not pictures of the real world – the instrumentalist position is clearly expressed here.

Paradigm / Approach	Critical Rationalism		Interpretivism	Complexity Thinking
Vision of reality (Ontology)	**Empirical Realism** An independent reality exists that we can access directly.	**Agnostic attitude** What exists outside our observation, in reality, we will never know.	**Constructionism** Reality does not exist independently of us; it is at least partly a construct of the human mind.	**Critical Realism** We have ample reason to believe that an independent reality exists. But which part of reality is objectively real and which part is constructed can never be determined exactly. Unmediated access to that independent reality is impossible.
Theory on how we can get to know reality (Epistemology)	**Objectivism** We can obtain objective knowledge of reality through logical reasoning and empirical research.	**Instrumentalism** Theories are tools to make predictions and instruments to test acquired knowledge.	**Perspectivism and Relativism** Our knowledge of reality is based on different and mostly incomparable conceptual schemes.	**Perspectivism and Fallibilism** Our knowledge of reality is determined by the different conceptual schemes we use. By comparing them we can try to obtain knowledge though never 100 percent objective, certain knowledge.
Definition of truth	Truth is defined by correspon-dence or isomorphy/ similarity between claims or models of reality and reality itself → 'objectivity'.	None. At most: 'truth' is what works. Theories are only viewed in terms of their usefulness.	At most, truth can be defined as coherence within frameworks of meaning, in overarching systems of knowledge claims about reality further → 'subjectivity'.	From truth as representation to 'truth' = what works via a network model of correspondence and coherence plus pragmatic criteria on the basis of critical intersubjectivity.
Analogy	Mastermind. Knowledge as representation of reality.	Science = tool.	Narrative / Story. Knowledge as narrative report or compass.	Mapmaking or Coral Reef. Science as tool and placeholder.
Danger	Apparent rock-bottom basis.	'Foul ball'.	Drift sand.	Misfit / 'Foul Ball' / Illusion of Representation (or no grip whatsoever).
Quality criteria / Ethics	Objectivity on the basis of claimed neutrality and value freedom with regard to the research object.		Immersion in the social world produces knowledge that might be subjective but nevertheless hopefully entails objective valid truth claims.	Objectivity on the basis of critical intersubjectivity and accountability with regard to held assumptions and own position.

Table 3.1   Complexity Thinking as a Potentially Integrative Meta-Position

Once again, the emphasis is on the notion that all discovered phenomena are the result of our interaction with the world we live in. Our theories and models are 'placeholders'; they are temporary substitutes that enable us to develop an increasingly complex understanding of the world. This allows us to 'negotiate' a reality that our intervention has made ever more complex. We can never catch up with that reality because every time we get close, we complicate the situation even further. Acquiring more knowledge never solves the problem; it creates new problems that we then have to find solutions to.

A metaphor that helps to explain this new way of thinking is that of a coral reef. A coral reef is nurtured and formed by the surrounding world: by sunlight, ocean currents, nutrients, other plants and animals, etc. To survive and grow, a reef has to adapt and interact effectively with its surroundings. Rather than represent its surrounding world objectively, it expands in relation to this world. And in the course of its evolution it develops and adapts to the changing circumstances, becoming more complex or becoming extinct. Knowledge acquisition follows a similar path, argue Julie Thompson Klein (2004) and Angus McMurtry & Jennifer Dellner (2014). As they see knowledge in terms of continually expanding viable relations, they think the metaphor of the coral reef is more adequate than an image of knowledge moving ever closer to a final destination.

Using the vocabulary of complexity thinking, we could say that knowledge emerges from our interaction with our surroundings and links back to those same surroundings. So knowledge and reality are not separate systems to be brought into line somehow. We should rather see them as part of the same developing, emergent complex system that is never complete at any given point in time. This is because complex systems are in a constant process of development, without ever reaching ultimate completion; they will never enter into a state in which they are 'as it is'. The advantage of this concept of knowledge is that it enables us to propose a theory of emerging knowledge in which the real world and our knowledge of it are both part of the same complex system. They are not two separate systems; they are inextricably linked.

This metaphor does not resolve the dilemma of realism versus constructivism by taking one particular side. Neither is it an attempt to find a middle road. Its purpose is to rise above the contradictory alternatives and to move beyond the opposing paradigms. To imagine knowledge as a coral reef, argue the proponents, does more justice to the complexity and interdisciplinary nature of knowledge about 'wicked' problems than the linear vision of knowledge as a foundation, or knowledge as a model or a cognitive map with contours and boundaries.

The view of the knowledge process as a coral reef certainly has some inspirational aspects. A big advantage of the metaphor is that it fits in better with an action-focused view of the concept of truth: it shifts the emphasis from attempting to discover a static and essentially unattainable truth to a dynamic process of 'making things come true'. Seeing the knowledge process as a matter of co-production – of interaction between those who know (epistemology) and that which is known (*metaphysics*) – the attention shifts from fact-making to sense-making (Jasanoff 2004: 274-276). It fits well with the aim of attaining 'know-how' (via modelling and design) as a counterpart to the 'know-that' reflected in knowledge claims that we take to be true or models of reality that we see as representative (cf. Olaya 2014). Theory becomes a form of practice: we change the world by understanding it, and we understand it by changing it (Byrne 2002).

One reason we should remain sceptical of this alternative is the risk of falling into the instrumentalist trap, since it raises the issue of how to decide whether we are successful in 'making things come true', and indeed whether or not a theory or model actually 'works' or not. This seems a high price to pay to avoid the danger of the distorting image of models and theories as representations of reality – a distortion that was already dealt with by exchanging the Mastermind metaphor for mapmaking. The conception of models and theories as maps of reality also enables us to incorporate the constructivist aspect of critical realism. And it seems equally compatible with its description as a 'placeholder', as temporary substitute that allows us to construct a more complex understanding of reality.

To avoid the instrumentalist trap, it is crucial to accept that we can only gain knowledge about systems that have been demarcated in some way, i.e. around which some boundaries have been drawn (Midgley 2000; Cilliers 2001, 2002). Only then can we escape the pitfall of knowledge acquisition being postponed forever (since we can never fully comprehend emergent complex systems). This is, in fact, the strategy proposed in the mapmaking metaphor, because we can acquire knowledge by creating a demarcated, simplified map of the actual situation or system. The best way out of the dilemma may therefore be to reserve the mapmaking metaphor for concrete scientific research projects while using the coral reef metaphor to visualise the overall knowledge process that transcends time, place and context – a process that never ends.

3.4 Towards a More Dynamic View of Science

Chapter 2 showed that it is difficult, if not impossible, to maintain the idea that the knowledge process is research by a neutral, indifferent knowledge subject (researcher) into a purely reactive, non-interactive knowledge object (phenomena being researched). The norms, values and power relations entailed in phenomena being researched cannot be discarded; they need to be taken into account in research. For many scientists, this is reason enough to extend the knowledge process beyond 'merely' finding explanations. They think scientists should actively commit to the urgent issues that society needs to address and use their knowledge to help find the best solutions for these problems. In this final section, we examine two scientific approaches that are particularly geared towards helping to realise change in society.

In essence, the empirical cycle focuses on finding explanations for knowledge questions, while the hermeneutic cycle focuses on understanding human action. This offers important insights into fundamental causal and interpretative processes which in turn can be used to figure out how we can tackle particular problems. Yet that is not the principal focus of knowledge acquisition, and this is something a number of researchers find problematic. They argue that rather than just provide explanations and interpretations, scientists should participate in developing new insights with which to solve the problems society faces. In the late twentieth century, a group of scientists began to develop a research method designed to search for and test targeted interventions to deal with urgent problems (see Coenen 1987; Denzin & Lincoln 1994; Greenwood & Levin 1998; Reason & Bradbury 2001). Their focus is therefore more solution-oriented.

The research cycle they propose is outlined in figure 3.3. This type of cycle is characteristic of action research, intervention research or evaluation research. Just as in the empirical cycle, the problem diagnosis phase relies on empirical observations, with the difference that in this type of research the fact that all observations are dependent on interpretation is given its proper due. Then, in the next step, a theory is developed about how to resolve the problem (compare this with the theory formation phase in the empirical cycle). Based on this, a concrete strategy is developed or an intervention is designed to solve a problem in a specific context.

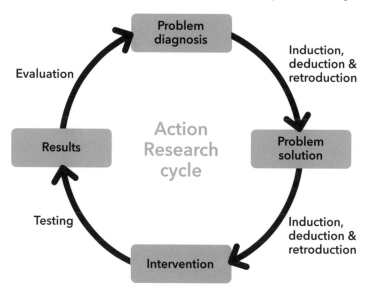

Figure 3.3 A research cycle with a focus on the action component

The way this intervention is tested resembles the experimental testing of hypotheses in an empirical cycle. This is followed by a phase in which the results are evaluated, which may lead to another, subsequent action cycle that builds upon the findings and results of the first one.

Action Cycle, Empirical Cycle and Hermeneutic Cycle in Food Research

Let us examine how an action cycle works in research on the food issue, and the possible role of input from the empirical and hermeneutic cycles in such an action cycle.

Research organised according to the standard scientific procedure (the empirical cycle) can help develop insights into underlying processes in crop growth and food production, the causes of disease, the role of nutrients and genetic processes. Since the rise of modern science, continuing research on the genetic composition of crops has given us a deeper understanding of how we can enhance seeds. Moreover, it has helped us to develop high-yield varieties that are crucially important if we are to feed the growing world population.

The question is how to use this knowledge in practice to develop successful solutions to the food issue. This may be through demonstration, a kind of research that takes the form of an action cycle. While the empirical cycle focuses on producing knowledge, the action cycle targets the testing of interventions to resolve problems in society - problems that are often unique to specific areas, with particular characteristics linked to a given location.

While experiments are held in the controlled situation of a closed system, demonstrations are about testing whether something that works in a closed system also works in the world outside. In the latter situation - i.e. in an open system - factors may come into play that cannot be maintained at a constant level.

To ensure that the results achieved in experiments are also achieved in demonstrations, it is sensible to test not only in the research laboratory but in the field as well. The action cycle can play a key role here; it interferes in the process (in the open system) and focuses on the search for a genuine solution to a concrete problem. Particularly in areas such as agronomy, where the problem is often to find solutions to food production in a specific climate with a specific type of soil and specific available nutrients, this is a fruitful method. Subsequently, the empirical cycle can be employed to test the advantages and disadvantages of promising approaches.

The action cycle can produce fruitful results in developing countries, for example, where the innovations employed elsewhere have not yet been successfully implemented. One way to stimulate this is by applying a theory developed in closed experimental settings in open field experiments in those areas. These field experiments can be crucial in persuading users – i.e. croppers – that the new knowledge is useful.

Figure 3.4 Research in a closed system   Figure 3.5 Research in an open system

The action cycle offers a way of finding solutions to problems identified in the empirical cycle. Part of the solution may also be found by closing the gap between the research and dissemination of the research results, because besides the actual research (with its fundamental and practical focus), communication and interaction with potential users is of crucial importance. Here the hermeneutic cycle comes into play again, because varying interpretations will play a major role in this communication and interaction. As a strict separation between research and the transfer and application of its results can lead to all kinds of undesirable problems (see chapter 4), some researchers argue that these functions should be integrated into agricultural research (see e.g. Fresco 2009; Maat 2011).

3.4.2 Design Thinking and the Policy Cycle

Since the arrival of the computational methods, new opportunities have arisen to develop actionable insights. Computer-assisted reasoning, for instance, enables us to design interactive digital visualisations that can help users generate hypotheses about what they think are the best strategies. We can let the computer test these hypotheses across an ensemble of plausible models, thus enhancing our ability to deal with problems characterised by deep uncertainty (Bankes et al. 2001: 71; DeTombe 1994, 2015). Combining models and design thinking can have a powerful impact when modifying and transforming real-life systems, particularly when the model creation process is linked to a policy cycle to produce creative solutions and to develop and implement refined policy strategies.

After the problem has been defined and diagnosed, hypotheses can be formulated about potential solutions and fruitful interventions, as in an action cycle. If the simulative runs indicate that the designed interventions are sufficiently efficient, they can be implemented in the real-life situation. Policy interventions are then tested, monitored and evaluated for intended and unintended consequences. If this shows that modifications are needed, the policy strategies can be adjusted. This follow-up to the original problem definition leads to a reconsideration of the diagnosis and adjustment of the proposed interventions, which can then be implemented anew (see figure 3.6).

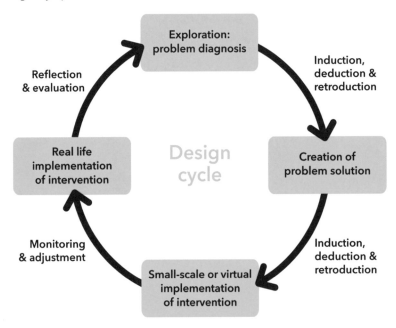

Figure 3.6   A design cycle with a focus on implementing solutions

For Byrne (2014: 118-119), the significance of complexity thinking lies in the recognition that

> 'whilst there is no inevitable outcome, no linear law, no single answer, we can nonetheless analyse in order to see what the possible set of outcomes might be, what the possible answers are, and [...] intervene in order to achieve those we want to see happen.'

There is not an infinite number of competing truths about the complex issues we study. There is truth, albeit a complex truth, Byrne argues. Our task is to work out how policy changes will affect the future trajectory of systems (Byrne discusses urban systems). By showing the range of possibilities, systems and design thinking can help us figure out how to prepare systems for their tasks in the conditions in which they operate. Thus, applied and grounded speculation based on simulation models is used to identify the historical dynamic of particular socio-spatial systems and to make forward projections. This is a new kind of complexity-based 'social engineering science', a rational programme not of declarations based on absolute predictions but of social action based on the specification of many – but not an infinite number of – options (ibid.: 146-167).

Social engineering is an expression that may raise some eyebrows in the social sciences and the humanities. We would definitely not want to welcome a type of complexity thinking that suffers from the same illusion of manageability as the traditional approach. However, the emphasis on agency here is a reassuring and inspiring aspect. The action cycle and the design-based policy cycle are fruitful additions to the empirical, hermeneutical and modelling cycles. They fill the gap with respect to agency, which is either left out completely in the empirical and hermeneutical cycle or is only accommodated in a weakened form, as in agent-based models. Steven Bankes et al. (2001: 76) provide an inspiring way to look at it: they argue that the combination of computer-assisted reasoning and human reasoning should be seen as a beneficial, symbiotic, problem-solving strategy where computers are used to track through the implications of enormous sets of data, facts and suppositions, while humans use their abilities in pattern recognition and abstraction to bring into view the bigger, overall picture.

At the same time, action and intervention research is susceptible to objections encountered in the interpretative approach: the validity of knowledge claims proposed in this type of research is a sensitive issue. There is also a risk of creating self-fulfilling prophecies, though these can also have beneficial effects if our aim is to make our strategies come true. In any case, if we are to tackle complex problems, we must employ action-focused types of research in our methodological portfolio.

While Byrne concentrates on the social sciences, his message is relevant to all kinds of disciplines involved with complex problems. We need to develop knowledge about physical structures and ecological systems, but we need proposals for action strategies to tackle the challenges as well. Chapter 4 takes us a step further in this direction by looking more closely at science's function in society and at the role it has actually played up until now.

Questions:

- What is the difference between the concepts of truth, verisimilitude and reliability?

- In what way is perspectivism the counterpart of empirical realism?

- What new view of the role of science does fallibilism offer?

- How does instrumentalism differ from empirical realism?

- What is the correspondence theory of truth? And what is the coherence theory of truth? To what extent can correspondence theory and coherence theory be reconciled?

- Why does Mastermind fail as a metaphor for an integrated approach to understanding reality?

- How does the mapmaking metaphor differ from the Mastermind metaphor?

- How does the mapmaking metaphor resolve philosophical issues relating to the Mastermind metaphor?

- Why does fallibilism not necessarily lead to relativism or even scepticism?

- What are the two most influential approaches in modern science?

- How can we build bridges between these two different and often contradictory approaches?

- What problem faced by interdisciplinary researchers does the concept of duality of structure offer a solution to?

- What does the stratification of nature imply?

- Define two types of causality and describe how they relate to different levels that can be distinguished in nature.

- How does critical realism relate to empirical realism?

- Which metaphor do you consider most appropriate for the process of knowledge acquisition? Explain why.

- What value does the action cycle add compared to the empirical cycle and the hermeneutical cycle?

- What value does design thinking add compared to systems thinking?

4 Science as a Rational Process

In the first three chapters, we have come to regard modern science as a systematic learning process through which we try to find rational explanations of and solutions to our problems. In this chapter, it is time to take a closer look at what this rational learning process actually entails and how it can serve society.

If we examine the underlying concept of rationality, we find out that the concept driving modern science was redefined over the last couple of centuries and has ended up being a much more confined understanding of what 'being rational' entails. This much more restricted concept of rationality has led to a society and knowledge system that may be very efficient and most certainly has contributed to progress and prosperity, but in other respects can be said to be sub-optimal. In modern society, the reduced, instrumental version of the original rationality concept has come to dominate to such an extent that the economic and technical interests seem to have taken over. This has resulted in a rather one-dimensional rationalisation process that does not always produce the outcomes we wish and that also leads to unintended, unwelcome side effects.

This gives cause for a critical analysis of whether and how science actually serves society. In this analysis, the presentation of science as an objective and neutral enterprise is scrutinised, however attractive this image may be for policymakers who prefer that their decisions be backed by seemingly 'hard' scientific evidence. In addition to pointing out the dangers of value-laden science and the cherry picking of data by policymakers and other stakeholders, we take some time to reflect on why it is that society often fails to absorb scientific knowledge. And we compare the traditional modus operandi of science with newer types of science that seem to hold more promise in this regard.

Having found out that some (if not most) of the 'wicked' problems we are faced with are unintended, unwanted side effects of well-intended scientific solutions, we are forced to conclude that rational decisions are not by definition wise decisions. Since we are not content to dismiss *modernity's* rationality process as a failure and regard it as a project that we had better abandon altogether, we investigate what we can do to bridge the existing gap between rational and wise decisions. This leads us to the conclusion that it is time to reconsider the limited, narrowed-down concept of rationality and replace it with a more comprehensive rationality concept, so that in the future we will be able to capture and assess all relevant aspects when exploring the value and feasibility of proposed solutions to our complex problems.

4.1 The 'Project of Reason'

Rationality is a crucial concept in the search for knowledge, certainly in the search for scientific knowledge. Science is, after all, a rational learning process. We look for rational scientific solutions to complex questions that confront society today. But to imagine there is only one specific concept of rationality would be misleading, for the meaning of this concept has changed considerably over time.

This began with the *Enlightenment,* a movement that started in the seventeenth century (see Kant 1784) and that entailed a disenchantment with and a demystification of the world. Considerations and interpretations of life based on mystical and religious worldviews derived from the Bible were increasingly pushed aside. What emerged in its place was the search for rational explanations for how the world works. As the Enlightenment progressed, a rationalisation process took shape in which only knowledge acquired through the senses and by logical reasoning were accepted as rational (compare with the two foundations of modern science: empiricism and rationalism). This rationalisation process coincided with the advent of modern science and the independent pursuit of scientific enterprise in academia as it has emerged throughout Europe in recent centuries.

Over the last two hundred years, the concept of rationality has gradually been narrowed down to its cognitive aspect (Bernstein 1983: 37; Toulmin 1990: 30-41) – i.e. the mental activities involved in learning, observing, thinking, interpreting and searching for solutions. Only explanations and solutions based on observation and experience obtained through the controlled use of the intellectual faculties were considered rational. As a result, the everyday logic of ordinary life came to be viewed as tainted. Responding by intuition (the unchecked use of our thinking abilities) and drawing conclusions based on singular experiences (inductive logic) were dismissed as invalid ways of acquiring true knowledge. Case studies, which involve research into specific examples that might provide interesting insights and experiences regarding wider social phenomena, were likewise discredited. These specific, personal and context-related experiences had once been considered a valuable source of knowledge. But at the end of the seventeenth century, the emphasis came to rest exclusively on timeless, context-free knowledge.

This rationalisation process lies at the heart of the concept of modernity. The pursuit of rationality can be viewed as a project that has at its core the search for absolute knowledge and that uses clarity, certainty and necessity as the main criteria to assess the acquired knowledge. The advent of this form of rationality laid the foundation for the rise of modernity; the development of scientific rationality was supposed to provide the groundwork for the progress of humanity. In the course of this process, the concept of rationality acquired a narrower definition. From a concept that encompassed what we learn from both theory and practice – from formal and everyday logic (common sense) – it was reduced to a definition that encompassed only elements regarded as 'pure' knowledge.

The principal cornerstones of this bastion of modernity are the division between mind and body, the notion that the world is governed by immutable laws, and the idea that emotions confuse rational thinking and must therefore be suppressed. These principles describe a system of ideas that provides the framework of modernity.

4.2 Unintended and Unwanted Consequences of the Rationality Process

4.2.1 Reduced and Reducing Rationality

At the very beginning of this book, we analysed how the many refined and intricate forms of technology with which we regulate our lives make modern Western society increasingly complex. This has led to an increase in specialisation, bureaucratisation and globalisation, which appears entirely rational since it benefits society's progress (Bell 1976; Habermas 1981a). Yet it is important to realise that the rationality of science, technology and economics is a particular kind of rationality, namely an instrumental one (Habermas 1981a, vol. I: 28). *Instrumental rationality* centres solely on achieving specific objectives efficiently. It is focused on the efficacious management and control of processes, and the necessary calculations and quantifications that come along with it. It can only accommodate procedures that lend themselves to formalisation and that conform to machine-like logic, i.e. procedures that work according to binary logic (yes/no, black/white) based on calculable and separable elements, just like computers do. The intelligence of an apparatus such as a computer is an extension of our own brain. But it is a restricted extension, as it is confined to the cognitive function, reduced to its calculating aspect (Lemaire 2010) and devoid of emotion, exactly as modernity's rationalisation project demands.

However, in the long run, this logic of the specialised systems feeds back to the users. The products and technologies we invented exert pressure on us to adjust to the instrumental rationality that we ourselves put into them in the first place. This leads us to structure our activities in such a way as to make them match the demands of computerised use, for instance by translating them in the only language they can handle: formalised language. Almost all processes related to administration, business management, taxes, banking, insurances, etc. are adjusted to the demands of computerised use. Time-saving mechanisms that have proved to be efficient in the business environment have been extended to other domains such as education, health care and ecological 'management'. There they can lead to questionable forms of restriction and impoverishment. All this is further enhanced by an economically focused society that turns almost everything into 'products' that can be financially valued and traded on the 'market' (see Habermas 1981a, particularly I: 455-534; Lemaire 2010: 53).

This 'rational' system that is synonomous with modernity has become more and more independent of the people who created and use the system and has started to lead its own life. Thus, humans may become unwittingly and unexpectedly controlled by their own system. However rational the system, the fact that we are being controlled by it inescapably contains an irrational element. Moreover, a rational system can easily be used for non-rational and ethically irresponsible motives, as history has taught us. Even the most rational aspects of society – science, technology and the economy – may at heart be driven by non-rational motives such as the desire to control, possess and manipulate (Habermas 1981a; see also Foucault 1979).

Thus, while the scientific enterprise has spawned a rationalisation process that has certainly advanced society, it has also had negative results in the form of unwanted side effects and risks (Beck 1986). If we want to find proper, sound solutions to the challenges that face us today, we will have to start by recognising what those unintended consequences are.

Unintended and Unwanted Side Effects of Our Agricultural System

The way that agricultural science is organised and the development of agricultural knowledge and methods have in many ways been extremely successful. At the same time, we cannot escape the conclusion that they have also caused certain structural problems. The methods and techniques proposed by science have resulted in drastic changes in the countryside, in the way farmers work, and in our own consumption patterns. These changes have solved problems related to agriculture and our food supply, but they have also generated new problems (including epidemics such as mad cow disease and Q fever) and exacerbated the imbalance between rich and poor countries as well as the social inequality between men and women. Moreover, new techniques created by modern science have sparked off a debate that continues to this day about the genetic modification of plants and animals (see e.g. Maat 2011; Fresco 2012, 2015).

In the case of genetically modified organisms (GMOs), there are, generally speaking, two camps of critics. There are those who warn about the dangers of genetic modification itself, while others may not disagree with the technology but have doubts about the potential abuse of GMO-related products by commercial multinationals. There are many NGOs in the former camp; they warn that by adopting GMOs, a weed may eventually develop that resists pesticides to which the genetically modified crop is immune. And they fear that this danger will grow the more agriculture relies on monocultures because this enables plants – including weeds – to become resistant to pesticides. This means that new strains of GMO plants must continually be developed to allow other pesticides to be used to which they are resistant (i.e. 'round-up ready'). Persistent weeds will also develop new resistant strains themselves, so that NGOs fear the problem will continually shift as so-called primary plagues (plant diseases and insect attacks directed against weakened crops) are replaced by secondary plagues (plant diseases and insect attacks against normal, vital crops). Meanwhile, they predict that the use of pesticides will continue to increase. Those in the second camp, which includes many scientists and also world leaders (see e.g. Haring 2013; Gore 2013: 261-269, 373), have no intrinsic objections to this form of gentech but find the practical results disappointing. Many share the NGOs' criticism that major players are motivated by profit and the desire to create a monopoly by patenting and are thereby obstructing the successful development of gentech.

It is hard to identify where the responsibility for these problems lies. Does it lie with science and the way in which innovations and solutions are offered without first properly thinking through the consequences? Or does the fault lie with the massive industrial agribusinesses and commercial multinationals and the way they use or even misuse scientific knowledge? Or could it perhaps be a combination of both?

4.2.2 Fragmentation and Alienation

One of the reasons so little attention has been paid to the consequences of GMOs is that growing complexity has made modern society less and less transparent. The domains of science, technology and economics in particular have expanded to become vast terrains with intricate mutual connections. Moreover, they are connected to all parts of the world – a prime example of how globalised our world has become (Giddens 1990, 1991). But how these domains work is often unclear for ordinary people. The functional and institutional differentiation in fields that have insulated themselves from society has led to a phenomenon that Jürgen Habermas calls the *fragmentation* of the lifeworld (1981a, vol. II: 521). Since significant areas of human life have become compartmentalised, society appears to individuals as a huge collection of more or less separate blocks, a multitude of fragments, making it difficult to see any cohesion between the various parts.

A broad range of practical specialisations and expert domains has developed; and these domains have in turn become increasingly insulated. New specialists are hence needed to open up these domains to other specialists or to the ordinary public – specialists such as communications personnel, managers and coordinators. The more experts there are in specific domains, the harder it is to oversee the connections and the unity of the whole. It is as if the more knowledge we gain and the more expertise we develop in a specific area or discipline, the harder it becomes to oversee the whole picture. This is the paradox of specialisation: increasing knowledge in specific domains seems in direct disproportion to our insight into the overall picture. We cannot see the forest for the trees.

Some philosophers also consider this to be true of our insight into the cohesion and unity of the world we live in. Processes such as rationalisation and the concomitant phenomena of bureaucratisation and fragmentation enhance the impact of abstractions in the life of the modern individual and can engender a sense of alienation (Giddens 1990, 1991; Habermas 1981a, vol. II: 513 ff.). We have created a world that is so complex that we are barely able to comprehend it. And the bigger the problems, the less we can influence or control them. Issues at the planetary level, such as climate change or the world food issue that is taken as central case in the book, are a striking and far-from-encouraging example here.

The social sciences have not escaped this trap of specialisation either, as the plethora of disciplines and therapies developed in this domain show (ibid.: 533). Even in philosophy, where there was a prolonged effort to remain focused on the overall picture, fragmentation has developed: the perception of reality has steadily splintered, leading to a critical stance towards generally accepted theories as well as scepticism towards sweeping claims of knowledge.

At the same time, traditional social connections have eroded; the core family can no longer be seen as the cornerstone of society and membership of social organisations such as unions has dropped. Consequently, the authority of certain institutions that used to be taken for granted – such as the church or political parties – are now open to discussion. The advantage of this is that people nowadays feel more free and mobile and that a greater diversity of subsystems and subcultures is permitted. Other major advantages of the rationalisation process are that everyone is formally equal and that everyone is able to demand their rights and may expect others to tolerate their life choices (Habermas 1981a, I: 465), in the western world at least. Yet there is also a downside.

Modern individuals need to process countless messages, words and images every day, which sometimes leaves them with a sense of fragmentation and emptiness rather than with enhanced insight and understanding. It is doubtful whether we are capable of dealing with the acceleration of time, the flood of information and impressions, and the complexity of a globalised world that modern society confronts us with. Stimuli and fragments of information received from near and far combine to form in the best case a rather disorganised perception and in the worst case a disoriented awareness of an incomprehensible universe (Lemaire 2010).

The disintegration of our social environment is reflected in individual experience. Modern individuals often have difficulty forming a coherent sense of identity. These days, people have more choice and freedom to connect with others and to find contexts in which to establish their sense of identity (Giddens 1991: 148-149). While they have become freer as a result of the loosening of family and social ties, they live a more isolated and abstract life. And they try to fulfil so many different functions and roles that they risk losing a sense of who they really are.

'*Life politics*' refers to decisions about life that arise from a process of self-realisation in a society in which people are no longer bound to traditional roles and expectations. An identity is no longer a given; it has to be continually maintained. It assumes a *reflexive* self-awareness of a person's own life story (ibid.: 1-9, 52-53). While liberation from established local social patterns and lifestyles may feel like a release, bearing personal responsibility for every step taken and every decision made can become a heavy burden for an individual and a source of fear and anxiety. It is no mean task to develop an inner authenticity, i.e. a framework for ideas about one's own identity and for the interpretations of needs to make sense of the course of one's own life. Especially if circumstances keep changing, it may become a daunting task to maintain a comprehensive, coherent narrative about one's individual life story.

This reflexive self-project takes place amid the multiplicity of choices that the individual faces in today's modern, abstract, global systems. This implies that people increasingly develop personal preferences, views and positions in a context in which social relations are detached from and elevated above a familiar personal context. Life politics is the politics of self-improvement, in which individuals continually test their standpoints against all kinds of abstract systems (ibid.: 214-215). The scientific expert system is one such abstract system, as is the legal system, the stock market or the World Wide Web. The combination of the fragmentation of one's lifeworld and the abstraction ensuing from globalisation can arouse a sense of alienation within modern individuals.

Many philosophers and social theorists have therefore concluded that modern society is based on a calculated or calculating sense of rationality (Toulmin 1990; Habermas 1981a; see also Beck 1986; Beck, Giddens & Lash 1994; Nussbaum 1990, 2015; Flyvbjerg 2001). Scientific rationality is not in itself the problem. What is problematic is that the modern, narrow version of the original concept of rationality – cognitive-instrumental rationality – has become so dominant that economic and technical interests now seem to have invaded every domain and become overpowering.

Thus there are two sides to the ongoing rationalisation process: progress and regression. While it may enhance technical and work processes so that they are fast, efficient and uniform, it negates other valuable ways to regulate our interaction with each other and with the world of nature (Habermas 1981a, vol. I: 102). Instrumental rationality serves only some of the interests of a section of the population, certainly not everyone. And it certainly does not by definition benefit all of humanity.

Moreover, with its increasing sociotechnical interconnectedness, today's society is more prone to uncontrollable cascade effects. This lowers our social system's resilience to stress and amplifies problems from the local level to a more global scale (e.g. Folke et al. 2010; Lane 2011; Helbing 2013;Törnberg 2017: 65-66). Some believe this has brought about increasingly severe and interconnected crises such as the climate crisis, biodiversity crisis, refugee crisis and financial crisis. Adequate responses from the political domain have been lacking, for politicians have not managed to reach consensus on a common direction, nor have they succeeded in propelling action (e.g. Leach et al. 2010; Loorbach 2010; Haasnoot et al. 2013). This should come as no surprise, considering the tendency within the political domain to emphasise prediction, planning and control. This strategy is far from appropriate in the current situation, which entails untangling socio-eco-technical webs of feedback and threshold effects, intertwined drivers and deep uncertainty (ibid.: 73). Increasingly, people have been arguing that this 'crisis of crises' (Beddoe et al. 2009, Lane et al. 2011) calls for a fundamental transformation in how we organise our society – what some call a Great Transition (Raskin et al. 2002).

4.3   The Societal Value of Science

4.3.1   Science as a Quasi-Neutral Solution Factory

Critical rationalism has developed into the most influential conception of science. This can be attributed to its service both as a scientific method to find explanations and its key role in stabilising the existing societal order (Kunneman 1986: 19; Dehue 1990: 216). Bureaucratic organisations and institutions prefer their policies to be supported by 'genuine' science. The critical rationalist model of knowledge acquisition, which is generally seen as the standard model of science, is well-suited to this ideal image. It is supposed to solely imply a view of the way scientific research should be pursued. It pretends to be able to produce objective and unprejudiced knowledge and maintains that scientific researchers have no particular interests that could lead to a biased perspective. Such a presentation of science ignores the real consequences that this methodological viewpoint has on our vision of society, and specifically for our vision of the role of science in society (Jasanoff 2012:14).

This same danger faces the latest scientific approach, in which computer simulations, equations, visualisations and other types of models are used that have a veneer of objectivity but nevertheless build upon parameters and assumptions that involve subjective choices and estimations. A model is never made without a group of users having a particular purpose for it. In formal models, normative decisions of framing are concealed by enciphering them in technical code. As long as they are undisputed, they help to maintain the status quo, including the hegemony of standard scientific methods. The claim that only epistemic values are implied in scientific models and methods is, however, severely disputed within the philosophy of science (see Winsberg 2010: 93 ff; Shackly et al. 1996: 203). The choices implied in boundary work, problem definition and even model parameters inevitably imply the selection of certain social and political preferences.

Harry Kunneman (1986: 21 ff) uses the concept of a 'truth funnel' as a metaphor to show that regular scientific problem-solving patterns in Western societies act as a giant fyke. The truth funnel metaphor shows how knowledge is filtered in science until only apparently objective knowledge remains. By providing an 'objective' description, all normative, expressive and aesthetic aspects are negated and only a seemingly 'neutral' and 'disinterested' description remains. This remaining description is merely a partial – or a funnelled – description in which the knowledge that is found is reduced to so-called 'objective' facts. Qualitative assessments, which are often considered to be vague, are not taken into account.

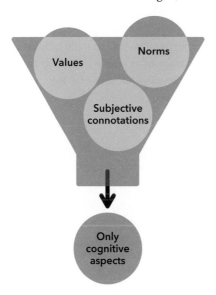

Figure 4.1 The truth funnel

The problem with the funnel rationality is that by selecting and representing only particular parts in the description, other parts are ignored. This means that the issue is no longer viewed in all its complexity. The 'objective' description also exudes a certain power: by presenting an explanation or solution as a complete, true description, policymakers are more inclined to base their decisions and actions on it. Describing problems in this way makes them easier to deal with. It is far easier to formulate narrow, rational, scientifically responsible solutions to a problem reduced to a few objectifiable aspects than for 'messy', complex issues linked to all kinds of intricate normative and expressive questions. Yet these supposedly 'objective' problem formulations are actually based on certain unrevealed values that nevertheless have an unmistakeable impact on society.

In fact, the standard scientific approach is characterised by a paradox: although science claims to remain above politics and politics claims to stay out of science, science is presented as the sole guarantee of rationality in political practice. Here we see once again that rationality is explicitly related to knowledge – specifically cognitive knowledge – and not related to norms, values or power, except in the indirect sense that rationality is itself seen as the greatest good (Habermas 1973: 243-244, 269; Van Peperstraten 1993: 206-208). The leading idea is that knowledge is equal to what science has to offer (this is called *scientism*); an opinion that is extremely limited and that only reinforces the narrow conception of rationality. It has led Thomas McCarthy (1978: 40-41) to state that belief in the sciences has itself become a dogma. In his mind, the adherents of scienticism are attempting to make science immune to philosophy by shifting problems related to the subjective conditions of knowledge to the domains of the psychology and sociology of science and by discarding reflection on underlying presuppositions as outdated metaphysical, or empirically unverifiable, thinking.

It is this pretence of being objective and value-free that makes it difficult to describe and criticise science's actual role in the continuation of existing power relations. And yet the particular approach chosen can have major implications for the structure and results of a research (see the example in the box).

The Controversiality of Scientific Research

Which scientific approach is considered best has a huge impact on what solutions are considered appropriate for a complex problem such as the food issue. And which approach is considered adequate depends in turn on the particular perspective taken as the point of departure.

Some researchers argue that a proper approach to the world's food issue should combine knowledge obtained from agro-ecology, agricultural studies and breeding technologies such as advanced chemical pesticides and the use of genetic manipulation. All of these approaches have to be based on scientific research, of course.

Many NGOs understand the importance of this underlying scientific research, but they question the claim that agricultural science and genetic technology can produce rational solutions to the food problem. They also have reservations about whose interests are being served with such an approach. Greenpeace, for example, may not oppose laboratory research into the underlying processes and potential effects of genetic technology, yet they do not see gentech as a solution to today's problems. They do not believe that potato blight can be solved by inserting resistant genes into a potato. They view the disease as the undesired consequence of today's agricultural system, and they argue that changing the system and departing from monocultural agriculture would be a better approach (Van Bekkem & Lotz 2013:11). Greenpeace wants a system change, a paradigm shift. Instead of 'end-of-pipe' solutions such as GMOs in the form of vitamin-A-enhanced rice or pesticide-resistant varieties, they prefer to address the food issue at its core and to change the basic structure of the food system.

Certain stakeholders sometimes cast doubt on the validity of research on which other stakeholders base their scientific arguments. Greenpeace

(2015) notes that the deployment of resistance devices is often based on research mainly carried out by major multinationals involved in GMO technology such as Monsanto. They consider it debatable whether a stakeholder can produce independent research while it is so clearly interested in a particular outcome of the research. The same is true of the circles in which much of agricultural research takes place. While close links bind together the government (e.g. the ministry of agriculture or economic affairs), universities and the agricultural sector, it can compromise the independence of researchers. Politicians who use or completely ignore scientific evidence at their own convenience to benefit key interest groups can be regarded as guilty of policy-based evidence picking (Martinuzzi & Sedlaçko 2016: 44).

But then again, NGOs such as Greenpeace are also accused of the selective use of available data. Louise Fresco (2015: 437), for instance, criticises their tendency to ignore inconvenient facts and insights regarding GMOs or to (re)interpret them in such a way as to confirm their position. In her view, this current widespread tendency of cherry-picking data makes it difficult, if not impossible, to conduct a dialogue between opposing parties.

What can be said in favour of NGOs is that at least they are open regarding the underlying values behind their activities and strategies, while scientists and politicians usually adopt an attitude of self-proclaimed objectivity and value neutrality.

4.3.2   Mode 1 and Mode 2 Knowledge

Originally, academics claimed that scientists study research questions purely out of an interest in how the world works. This was considered a condition for conducting neutral and value-free science. Scientifically derived knowledge explained natural and later also social phenomena, and it was up to society to decide how to use that knowledge. This image of science has changed very little. True, while it is generally recognised that questions addressed by scientists are often inspired by technological or social issues, it is still maintained that knowledge gained through science is nevertheless objective. Science is considered an independent occupation in an independent position. Science serves society only in the sense that the knowledge gained will hopefully find its way into society and be of some benefit. This vision of science is called mode 1 knowledge acquisition (Gibbons et. al. 1994; Nowotny et al. 2001).

One characteristic of mode 1 knowledge production is that problems are defined and solved in a context determined by the mainly academic interests of a particular scientific community. Second, this form of knowledge production focuses on universally valid knowledge, which scientists try to find by following an approach that is as homogenous as possible (the standard scientific method). Third, mode 1 knowledge acquisition generally involves a single discipline; it focuses on finding explanations about how certain processes work in a particular environment – for example physics, chemistry or psychology. Thus, this approach is linked to the basic concepts, central assumptions and standard methods of these disciplines. Finally, mode 1 is characterised by a vertical, hierarchic relation to the environment: it is the academic community that determines what are relevant research questions and which answers suffice. The acquired knowledge and the methods employed are specially designed to solve these types of questions.

Mode 1 refers, in short, to the cognitive-instrumental knowledge that matches the familiar traditional, monodisciplinary forms of knowledge. It focuses on the production of supposedly objective, value-free, universally valid knowledge. The quality of the acquired knowledge is controlled by the scientific community itself, and it is left to the rest of the community to decide what to do with this knowledge.

Gibbons and Nowotny (ibid.) have identified a new form of knowledge production that has emerged since the turn of the century which they call mode 2 knowledge acquisition. Mode 2 knowledge acquisition takes as its point of departure practical, context-linked questions and problems. It is less about producing universally valid knowledge than about finding suitable solutions to practical questions and problems. Mode 2 is about realising things, making things come true; it can be viewed as transformational knowledge, as transactional problem-solving (cf. Schön 1987: 73, 79; Hirsch Hadorn et al. 2006).

In mode 2, knowledge is used in specific application contexts. Consequently, in mode 2, knowledge is both interdisciplinary and transdisciplinary. Compared to mode 1, mode 2 focuses more on society and is more reflexive, implying that researchers reflect more on the basic assumptions underlying their endeavours. It is therefore subject to a different kind of quality control. Its validity is relative to the researcher's commitment to a particular appreciative system and overarching theory (ibid.). Yet it is not only up to a specific scientific community to decide what is successful and what is not. Other stakeholders also exercise quality control. This primarily involves a wider, more heterogeneous and dynamic circle of scientists from different associated disciplines. Second, it involves circles of commissioners and financiers. And third, it also involves engaged citizens, NGOs and interest groups, including action groups or user organisations, who wish to have their say regarding the value of the acquired knowledge.

Since the mode 2 approaches to problem-solving are not confined to one particular academic discipline and indeed bridge the academic and non-academic world, they may be described as transdisciplinary research practices (Brown et al. 2010; Repko 2012: 20).

	Mode 1	Mode 2	Mode 3
Type of knowledge	**Mainly monodisciplinary, cognitive-instrumental knowledge** focused on finding explanations for our academic and technological questions, with an emphasis on bèta knowledge. Sometimes also some multi-, and interdisciplinary knowledge (bèta and gamma).	**Mono-/multi-/inter-/ transdisciplinary, normative and value laden knowledge** focused on finding solutions for our practical societal questions, implying both bèta and gamma knowledge. Often also transdisciplinary knowledge, including experiential knowledge of non-academic stakeholders.	**Trans- or 'sans'-disciplinary, moral and existential knowledge** focused on finding answers on our existential questions of life, drawing mainly from the alpha domain. It can be regarded as transdisciplinary when experiential knowledge intuitive insights and are involved. It is sans-disciplinary when it comes to wisdom gained by life experience.
Relation Science / Society	**Separate domains / science = knowledge producer** Scientific knowledge is produced in a detached domain, separate from society. There is some interaction, but this only involves some tuning – there is no cooperation.	**Co-production / co-creation** Science and society are both active to find solutions for complex problems. Where needed, they cooperate to implement changes and monitor the effects.	**Guiding / inspiring** Moral learning processes result in value frameworks that could give direction to scientific knowledge development and technology implementation in society. But this interaction is often lacking.
Assumed role of scientific knowledge (development)	**Autonomous / instrumental** More scientific knowledge leads to societal progress. The development of policy related and/or applied knowledge will lead to the solution of societal problems and stimulate the economy.	**Interactive / service oriented** Scientific knowledge forms part of the collaborative development and problem solving process. In turn, this co-creative research process forms part of scientific knowledge development.	**Normative / directive** Since scientific knowledge doesn't only lead to desired, beneficial outcomes but also unintended side-effects, it is important to reflect on the effects, as well as on whose interests are being served by it.

Table 4.1   Different modes of knowledge acquisition
Based on Gibbons & Nowotny 1994; Thompson Klein 2004; Kunneman 2005; Regeer & Bunders 2007: 12

In this new vision, scientific research is said to be comparable to a business enterprise. It is not an isolated search for knowledge in an 'ivory tower'; it is an endeavour, a practice alongside many other social practices. It is an endeavour that generally involves all kinds of technical and social questions and that is related to the interests of government, industry, companies and social interest groups. Gibbons and Nowotny (1994, 2001) explain that this new form of knowledge came to dominate scientific knowledge acquisition at the close of the twentieth century.

Building on mode 1, mode 2 is still tied to the material possibilities and constraints that have been clarified by mode 1 knowledge production. Yet within the space discovered in mode 1, mode 2 forges its own paths, leading to the development of new knowledge. In mode 2, it is not just about applying knowledge acquired in mode 1. It is about developing knowledge in areas where new experiments are undertaken, using new research equipment to find answers to specific questions in practical contexts. Whether mode 2 results are further developed and produced on a wider scale basically depends on economic and political priorities.

Mode 1 and Mode 2 Knowledge in Agriculture

Agronomy has brought forward many insights and methods that have helped to increase food production considerably. Just consider the knowledge gained in the last two centuries about cause-and-effect relations in heredity, inflorescence, the effects of fertilizers and irrigation, and the fight against disease and plagues. These causal explanations – characteristic of the mode 1 approach – gave an enormous boost to our crop breeding programmes. The underlying aim to produce universal knowledge is also typical for a mode 1 approach. The hope was that by combining plant material from various areas, a variety of crops could be bred that would do well everywhere. The promise that GMOs will lead to higher yields often also entails a general claim, i.e. a claim that is supposed to hold true in all circumstances.

Another aspect that is typical of the mode 1 approach is the idea that science has a relatively autonomous position within society. Science is presented as a facilitating factor: agronomy provides the knowledge to optimally connect various elements in the food chain, such as the availability of land and water for food production, demand, and transport. In this way, it hopes to serve the process of steady progress in the food domain. Guaranteeing a good connection between science and society and enhancing the application of knowledge is not directly considered a task for science. Nowadays, there is the obligatory paragraph on 'dissemination' in proposal forms, but scientific researchers will not involve themselves too much with how their knowledge is implemented in concrete practices.

Publications on the results of scientific experiments are mainly written for and accessible to colleagues in the academic field.

Nevertheless, the continual developments in agronomy have had quite an impact on what is produced and how. The new techniques and methods that have been invented over time have led to fundamental changes in the way farmers and croppers work as well as in our dietary patterns.

It is clear that scientific knowledge does eventually find its way into societal practices. At the same time, the transfer of scientific knowledge does not always work out the way it is supposed to. When general theoretical insights are applied in a specific, concrete situation, often these insights turn out to be less universally valid than expected (see Maat 2011; Fresco 2009, 2012: 13, 37, 479).

Experimental fields are the principal instrument used by researchers to test theories developed in the confined space of the laboratory in an open system. Needless to say, the conditions in experimental fields differ from those in the laboratory, as the open system cannot be controlled in the same way as the closed system of the lab. Crop growth in the open field depends on numerous factors, many of which vary considerably between and during seasons. If a crop fails to grow as anticipated, it may be difficult to determine the cause. This area of experimental research in the field is where the connection is made between the theory found in the laboratory and the concrete practice in specific contexts.

Field experiments can be crucial in convincing local users that the acquired knowledge is fruitful. To be able to relate to user's specific regional and local circumstances, field experimenters need to develop substantial knowledge of the various kinds of farming. This knowledge has been acquired irregularly since the 1960s, outside or in the margins of agricultural science. Researchers examining specific contexts (open systems) need additional knowledge and skills, such as knowledge of specific soils and ecological circumstances and of the different types of agriculture as well as the different types of agricultural organisations (Maat 2011; Fresco 2012: 409). This is, in effect, an argument for using a mode 2 approach, which emphasises the relationship of theoretical knowledge to practice and which recognises the value of the knowledge that users on site can add to the academic knowledge of scientific researchers.

Nonetheless, certain mechanisms encourage scientists to prefer a mode 1 approach. One of these mechanisms is the constant pressure on research institutions to produce new, interesting solutions. The promises of a breakthrough based on new insights gained from studying genetics and GMOs have been especially attractive in this respect. It is not hard to understand why this is, considering the economic benefits attached to being the first one to find a new solution: e.g. chemical companies producing artificial fertilizers and pesticides will have a competitive advantage when they can claim, based on genetic modification research, that the profit from harvests will be bigger. Mode 1 science is particularly designed to find new insights that hold promising solutions, whereas the tiresome business of implementing the solutions in practice and trying to find out if it really works in open systems is left to those active in mode 2. Moreover, scientists often prefer to do research under optimal controllable conditions to increase their chance of publishing in a prominent journal, since that is the ultimate measure of success for academic scientists. A mode 2 approach employing experiments outside of the laboratory in an open system is less likely to produce publishable research.

In practice, there is often close cooperation and coordination between science and business, in this case between agronomy and the chemical sector. A good example of this kind of entanglement is the GMO product known as golden rice, a variety that contains extra nutrients introduced by genetic modification. The development of the rice strain was the result of a cooperation of agronomic scientists and a global chemical concern, Syngenta. The research was not directed solely by the researchers; non-academic stakeholders were also involved. This kind of cooperation is a mode 2 kind of science that differs from the scientific ideal of mode 1 yet it is still presented as if it is independent research. Such practices are problematic in so far as the public is often unaware of the entanglement of scientific and business interests. As long as the cooperation is not transparent, the organisations involved are not held accountable for what their research leads to, nor will anybody question whose interests are actually being served by it.

4.3.3 (How) Does Science Find its Way into Society?

Mode 1, the approach to knowledge acquisition that universities have traditionally focused on, produces highly valued knowledge. It is a key source of innovation in business, industry and society in general. But the usefulness of this type of knowledge is necessarily limited outside of the controlled environment of academic research. The pipeline perspective, which assumes that knowledge finds its own way from scientific research into practical applications outside the academic setting (see e.g. Laursen & Salter 2004: 2; OECD 2002), appears to be overly optimistic.

This lack of knowledge dispersal is neither the fault of universities nor the fault of companies – it is the result of the fundamental character of mode 1 knowledge production. It is important to realise that mode 1 produces a kind of knowledge that only applies to a limited section of the broad rationality spectrum. It is designed to generate cognitive-instrumental knowledge – facts and new technologies – and therefore avoids the normative aspects and power factors that are also inherent in every rational learning process (Habermas 1981a) but that are 'messier' to deal with.

Second, the mode 1 model is not designed for implementing knowledge. It focuses on producing knowledge for general application, leaving companies and society empty-handed when it comes to deciding on how best to use and implement the findings in practice (Gibbons et al. 1994; Nowotny et al. 2001). Mode 1 is in line with the vision of knowledge management that focuses on what facts are verifiable or falsifiable and what technical knowledge is available. It has little to say about the underlying process of knowledge management and does not provide an answer to the question of how knowledge can be disseminated in society or how it can be used for the general good (*valorisation*) or to help realise change (Tromp 2012: 18; see also Senge 2005).

The mode 2 approach is a departure from traditional science in that it does not take for granted that scientific knowledge will be disseminated and applied in society. Adherents of the mode 2 approach maintain that knowledge that aims to find solutions to complex social issues is acquired in a collective learning process. They see knowledge acquisition as the result of a process of co-experimentation and co-creation, of interaction between scientific researchers, members of the community and other stakeholders. This interaction is part of a dynamic, multi-layered learning process (Schön 1987: 153, 260; Gibbons & Nowotny 1994: 87-88; Nonaka 1994; Nicolopoulou 2011; Regeer & Bunders 2007; Byrne & Callaghan 2013: 208; Nowotny 2016: 106).

Transdisciplinarity is a collective name for all kinds of attempts at reflexive co-design, co-production, co-creation and co-evolution in science, technology and society. Interfaces are created in transdisciplinary research, bringing science and society together by working with specific social actors and interest groups, generating knowledge and solutions to urgent tangible problems (Bergmann et al. 2005: 15-19; Pohl & Hirsch 2007; Hirsch Hadorn et al. 2008; Godeman 2010: 629). Taking an actual question as the point of departure, a structure is designed to jointly develop robust knowledge that applies in theory and works in practice. In the next chapter, we will discuss this form of research in greater detail. Here we first address the question whether scientific insight leads by definition to the right decisions and whether in addition to knowledge it also provides a more profound wisdom.

4.4 Does Knowledge Also Imply Wisdom?

4.4.1 Towards a More Sensible Continuation of the Rationalisation Process

We have come to regard science as a systematic learning process in which we look for rational explanations and solutions to the complex problems facing society. The rationalisation process and instrumental rationality have definitely brought us further, especially in terms of mode 1 knowledge. But few of the utopian stories about the promise of rationality have materialised. Indeed, events have taken place in the last century – many in the name of rationalism – that fill the darkest pages of human history (think of the Nazi period in Europe, the Soviet Union under Stalin, the Chinese revolution under Mao that entailed everything but a 'Great Leap Forward', or the Cold War).

All this led many philosophers of the late twentieth century to compose the swan song of modern thought. The negative consequences of instrumental rationality gave them sufficient reason to bid farewell to the rationalisation project of modernity. They had lost their faith in the myth that humanity would develop ever further with the rationalisation process and thereby improve the world. Henceforth, they argued, we should focus on developing a postmodern culture, i.e. a culture that is no longer based on myths and one that abandons the belief in rational, linear development.

There are plenty of philosophers who feel we should not give up hope yet, precisely because of the one-sided, limited conception of rationality that has been used up until now. They argue that, despite its many unintended and unwanted consequences, we are only halfway into the modernity project. They reject the notion that modernity has become obsolete; some even claim that we have never been modern at all (Latour 1993, 2003, 2017) and that, in fact, today's society is really only *half-modern* (Beck 1986; Beck et al. 1994; see also Habermas 1981b). While societies have optimistically started to realise the ideals of liberty, equality and fraternity, the balance of the last three centuries shows that there is still a long way to go. The rationality project is still useful in their view, even if rationality requires a fundamentally different conception and the modernity project a new direction.

Alternative approaches are needed to develop rationality's potential to steer science in another direction. That does not mean throwing overboard everything that the old rationality concept stood for. It does mean creating a new balance in today's concept of rationality using the idea of reasonableness (cf. Toulmin, 2001: 44; Flyvbjerg 2001: 53-54). This insight aligns with the ideas of the Counter-Enlightenment, a movement developed by philosophers and social theorists of the so-called Critical School in Germany in response to the excesses of the Enlightenment. In contrast to the universalist ideal of Enlightenment reason, they argued for the importance of emotion and imagination, which they saw as an expression of individual awareness relating to a specific culture and a specific age (Leezenberg & De Vries 2005: 120). Many postmodernist thinkers have voiced similar opinions.

The new directions currently being taken in science can also help us continue the rest of this journey, particularly when complexity thinking is regarded as a programme of reason in understanding and of action informed by understanding, as Byrne (2014: 116) does. Recognising that challenging characteristics such as non-linearity, self-organisation and emergence render it impossible to design final, silver bullet solutions for our complex problems, Byrne nonetheless believes that we can come up with thorough analyses and well-founded rational interventions. Were he to assign complexity thinking to any of the positions, he claims it must be assigned to modernism rather than postmodernism. But if we are to carry forward modernisation successfully, it seems necessary to continue the rationality process in a different, less destructive way.

While the old concept of rationality was related to expressions of necessity and certainty, we now have to learn to live with what is reasonable, accepting the ambiguity and uncertainties that are inherently connected to knowledge acquisition. The universal validity of the criteria of rationality is an illusion; there are no permanent, uniquely authoritative principles of knowledge acquisition. What is rational is a question of optimising within given circumstances (Rescher, 1998: 168, Jasanoff, 2012: 19-20). So it is better to associate the idea of rationality with specific functions of human reasoning and to see how they can best be aligned. It is all about finding a good balance between theory and practice, between formal logic and our wider practical ability to reason (*rhetoric*) (Toulmin 1990, 2001; Flyvbjerg 2001).

Hopefully, this approach will also provide an antidote to the post-truth age that some claim we have entered (Davis 2017; d'Ancona 2017). It is a far closer match with the way people really think: partly using conscious arguments yet also driven by unconscious, moral principles and emotional motives (Lakoff 2010). Morin (2008) accordingly argues that the paradigm of complexity thinking should take account of the limitations of traditional logic and be combined or even integrated with other kinds of logical principles that we employ, such as dialogical and translogical principles.

So the challenge is to find the right balance between the hope of certainty and theoretical clarity and the impossibility of avoiding uncertainty and ambiguity in practice. This requires a revision of the narrow instrumental conception of rationality that has often proved to be incapable of producing decent, well-founded solutions. It may produce valuable knowledge, but it cannot tell us about what matters most in life: what are the most pressing questions, how best to apply our general scientific knowledge to improve people's quality of life locally, or how to treat our planet to be able to live a prosperous life in the future. The rationality concept needs to be extended again to counteract the distortion caused by the truth funnel. Instead of portraying science as a disinterested, neutral enterprise guaranteed to produce universally valid objective knowledge, the normative, expressive and aesthetic aspects that are filtered out in the traditional scientific approach should be made explicit. Only then can science's place in society and the role it fulfils there be critically examined.

It is exactly at this point that the distinction between different types of knowledge or knowledge production in mode 1 and mode 2 can prove its significance, for this distinction can help us acquire insight into the profound socialisation of scientific knowledge acquisition during the last decades. On the one hand, the spread of mode 2 has created more research directly focused on economic and political priorities. On the other hand, philosophers, theologians, ethicists, socially engaged scientists and other groups in society are trying to influence the direction in which science, with its primary focus on technological progress, is developing. They do this from a horizon of values that should be served by scientific research but that are often jeopardised by this same scientific research. To reflect on these values, we actually need a third form of inquiry – a form of inquiry that can help us to determine what is the wise thing to do when faced with complex issues (Kunneman 2005). This is where mode 3 enters the picture.

4.4.2 Slow Questions

Within today's modern knowledge society, the term 'knowledge' is constantly linked to the practical use of knowledge produced in mode 1 and to the efficient organisation of knowledge produced in mode 2. We must not forget, however, that existential questions and questions about moral issues that are also related to the development of a knowledge society require different kinds of insight and different kinds of learning processes. The 'head' – the cognitive-instrumental rationality that relies on purely formal logic – is distinct from the 'heart', with its broader ability to reason and to incorporate rhetoric and intuition (compare with the division between mind and body in Senge 2005: 157). When the head rules the heart, morality is disconnected from scientific thinking. That can be useful, since it makes it easier to apply science in quick solutions, thus making it more efficacious and profitable. This means that scientific progress does not need to stop and consider the moral implications of rapid developments and their evident impact. Many argue that this is how science has been able to achieve its present dominance (see e.g. Klukhuhn 2008: 326 & 362; Kunneman 2005), which is not without some serious social costs though.

It is against this background that Kunneman suggests an addition to mode 1 and mode 2 knowledge by introducing yet another kind of knowledge that he labels mode 3 (see Table 4.1). Mode 3 inquiry is focused on addressing 'slow questions'. This involves questions that are not just complex; they are so fundamental that we are unable to provide a definitive answer. Such questions include normative and moral aspects, e.g.: what are the essential characteristics of human nature? How should we relate to each other and the physical world in which we live? What is good and what is the best way to act in this respect? What is the point of living? And at what level of prosperity can the world carry on?

To pose these questions does not imply that we will receive a quick answer. But posing the question is a first, vital step in the process of formulating viable pathways to solutions. It would be wise for us to consider possible answers to these and similar questions, for these are questions that can point us in a particular direction when arguing for change in society. They are also a significant source of inspirational visions that are the driving forces of social change.

Moral Dilemmas Relating to the Food Issue

In the debate about what is the best strategy to tackle the global food problem, the proposal to increase food production by employing the genetic modification of organisms has provoked a vociferous discussion. A range of techniques is now available to optimally improve crops, for example by altering their genetic composition to increase yields with the right amount of nutrients and water. Besides the traditional techniques for the improvement of yields, a wide range of tools is available to employ bio and genetic technologies: tissue culture, DNA finger printing (identification based on short pieces of DNA), the option to mark genes, cloning, embryo transfers, restrictive enzymes (molecular scissors that sever DNA at specific places) and recombinant DNA techniques that use these methods to insert a new gene into a genome. Some see the latter - known as genetic modification - as just another step in the development of bio and genetic technology. They view it as a progressive line. Others argue that a boundary has been crossed: inserting a new gene into the genome of an organism is going a step too far. The mode 3 question here is whether humanity should be permitted to change the genes of other organisms and so alter their basic biological structure.

Considering the food issue from the distributive perspective, the question is whether prosperous countries with better research facilities are morally obliged to help solve the inequalities in the distribution of available food. This moral question can also be characterised as a mode 3 issue. The question is all the more relevant and urgent now that the number of people who have trouble dealing with abundance (people suffering from overweight and obesity) has surpassed the estimated number of people suffering from hunger.

The third source of knowledge refers to intuitive knowledge and focuses on exploring insights into moral questions and existential issues in light of the practical quandaries and dilemmas we face. Mode 3 learning processes tend not to follow a linear pattern but to pursue their own logic and depend on a certain friction between different views. Beside a variety of perspectives, mode 3 knowledge acquisition also employs inspiring narratives, visions and metaphors connected to moral dilemmas and major existential questions. This is the source we draw on when we face practical challenges and when we need to deal with the inevitable tensions associated with these challenges.

A wide range of resources is available in modern society to support mode 1 knowledge acquisition and to adapt and transform the acquired knowledge in mode 2. Kunneman (2005: 116) argues that such resources are lacking for mode 3 knowledge and insights. The conditions that would enable specific learning processes (including moral learning processes) are either barely fulfilled or completely lacking. The possibility of developing insights and sharing these through a 'creative tension' between various perspectives (ibid.: 126; Senge, 2005: 132, 140 &156) is, to a large extent, barred. This may be in the form of an appeal to restricted, dogmatically defended interpretive meaning frameworks or values. Relative and random values may also be defended in the name of individual autonomy or may simply be enforced by the more powerful or the smart (the latter may be seen as variations of the arbitrary or dogmatic stop in our thinking that we discussed in chapter 1). As a result, the development of science has in effect come to rest on an extremely limited interpretive meaning framework with respect to major existential questions and moral dilemmas.

In mode 2, various sorts of knowledge and insights are connected with each other to find adequate solutions for specific, context-bound questions and problems. Besides the objectivising knowledge that is employed in mode 1, we also have morally charged values that help determine the technical and organisational design processes we develop to achieve our practical goals (cf. the purposive, evaluative, pragmatic and normative aspects of transdisciplinary research identified by Hirsch Hadorn et al. 2006: 125 and Pohl & Hirsch Hadorn 2007: 118). Mode 3 knowledge acquisition focuses precisely on the normative character of these values as well as on the underlying interpretive meaning frameworks associated with the context-related questions and problems dealt with in mode 2.

These values and interpretive meaning frameworks should not be swept under the carpet (which can happen when they are filtered by a scientific truth funnel) but should be explicitly discussed and reflected upon (Kunneman 2005). That is the only way to make them transparent and part of the wider social learning process. It is questionable whether science is the best forum for this discussion. Perhaps reflection about the fundamental values and interpretive boundaries should be left to literature, religion or the arts (cf. Nussbaum 1990 & 2015).

4.5 From Funnel Rationality to a More Comprehensive Rationality

In modern Western societies, the domination of cognitive-instrumental rationality prevents the consideration of other aspects of rationality (Habermas 1981a: 102, 485). This is due to the tunnel vision that has led to a very narrow notion of rationality. Every problem is squeezed through a cognitive filter before it is studied 'scientifically'. And this approach continues to dominate to a large extent how and in which direction solutions are sought and found.

Naturally, our limited means and resources force us to make choices regarding the tools we use, or indeed the problems we choose to attempt to solve. Our choices reflect our perspectives regarding the world in which we live and how we relate to the natural world and to each other. To reflect critically on our attitudes and to take considered decisions regarding fundamental questions also requires rationality, but rationality in a wider sense.

Peter Senge (2005: 157-158) thinks that one of the primary contributions of systems thinking may prove to be that it helps to reintegrate reason and intuition. Indeed, the rejection of linear, non-systemic types of analyses may lead to a revision of the idea, so dominant in modern science, that rationality is opposed to intuition. Disregarding the question whether they ever were suitable, the formal, calculated rationality criteria certainly do not suffice as sole measures of the adequacy of knowledge today. This is because the challenges we face are not only about cognitive issues, they also relate to political matters and to moral and existential uncertainties. The old notion of rationality no longer matches today's demands of a modern society, which is more about diversity and adaptation than stability and uniformity. This is especially so since everything has a practical side that is evaluated on the basis of reasonableness (Toulmin 2001: 183-185). It is crucial to know the cause of a problem, for sure, but we also have to find ways to deal with the uncertainties and ambiguities involved in the problem analyses. And we need to reach agreement about the priorities of the problem agenda, about the values involved, and about the policy instruments we need to develop to solve the problem. To be able to do this, we have to make use of the critical thinking that drives scientific knowledge acquisition. But we also need to draw on the imaginative powers from other fields.

To tackle these questions in a truly rational way, we need to draw upon as much of our available and reliable knowledge as possible. How we propose to do that – to bring together knowledge from different disciplines – and how we propose to employ this to create robust solutions to urgent complex questions is the central topic of chapter 5.

Questions:

- What new form of rationality emerged in the late seventeenth century?

- This new form of rationality can be regarded as the foundation of modernity. What are the main basic ideas of modernity?

- What is meant by the fragmentation of the lifeworld? In what sense and at what level is this fragmentation experienced?

- Fragmentation may also lead to the development of specialisations. In this context, some speak of a 'trap of specialisation'? What does that imply?

- Toulmin and Habermas consider instrumental rationality a narrow version of the rationality that inspired the Enlightenment. What is the point of this instrumental rationality? What is its greatest disadvantage; what is the objection to the domination of instrumental rationality?

- What does Kunneman mean by the 'truth funnel' metaphor? What is the impact of this funnel rationality on knowledge production?

- Clinging to critical rationalism and the existence of objective value-free knowledge has been called a paradox. Explain the paradox involved here.

- Give four characteristics of the dominant view on scientific knowledge, i.e., mode 1 knowledge production.

- What are the characteristics of the newer form of mode 2 knowledge acquisition?

- What is valorisation? What barriers currently exist to bring the university's task in this regard to fruition?

- What, according to Klukhuhn, are the reasons that science has managed to acquire its current dominant position?

- Why does Kunneman consider it necessary to introduce a third mode of inquiry alongside mode 1 and mode 2? In your answer, make use of his critical analysis of the way the truth funnel or funnel rationality works.

- Can you think of an example of a 'slow question'?

- Which form of knowledge production (mode 1, mode 2, mode 3) accords with the modern rationality concept?

- And which form of knowledge production (mode 1, mode 2, mode 3) accords with complexity thinking?

- What is the connection between mode 1, mode 2 and mode 3 knowledge and multidisciplinarity, interdisciplinarity and transdisciplinarity?

5 Robust Knowledge for Complex Problems

Studying complex issues at the interface of humanity and the planet means that we need to relate to 'wicked' and 'messy' problems that are driven by intricate causal relations, correlations and complex feedback loops. Ensuing from constantly changing environments, they are characterised as value-laden, open, multi-dimensional, ambiguous, unstable, uncertain and unpredictable. Still, we would hope that science enables us to develop knowledge that is robust enough to help find solutions to these rather persistent issues. In this chapter we will find out whether and how this can be done by reflecting on the implications of complexity thinking for the scientific research practice.

We start with a short recapitulation of the various functions that the available types of research can perform. Then we review how they can be combined and integrated to meet the demands of complexity thinking, and what this requires of science. We devote special attention to the question of how research projects can be designed in such a way as to enhance the engagement of scientific researchers and other stakeholders in real-life complexity.

Next, we address whether and under what conditions we can maintain the claim that science leads to societal progress. Arguing that the traditional standards for scientific knowledge are not suited to assess the knowledge processes involved in inquiries into 'wicked' problems, we reflect on what could be regarded as more adequate quality criteria for present-day science.

In the conclusion, we evaluate what this all means for the institutional make-up of society and for researchers who are engaging in projects concerning 'wicked' problems. We summarise the types of knowledge they need to acquire and the kind of skills they need to develop to be able to deal with complexity.

5.1 Towards a Complexity-Based, Integrated Research Approach

In mode 1, the standard method is the leading model. Research projects are preferably set up as empirical or modelling cycles. Via systematic research, increasingly sophisticated theories are constructed from which hypotheses are deduced or projections and simulations are designed that are tested by empirical experimentation, model runs, statistical inference and mathematical computation. This way, scientists hope to be able to find suitable explanations for the underlying patterns and processes involved in 'wicked' problems. Academics make scientific claims and models, and it is up to politicians and policymakers to decide on the potential societal value of the generated knowledge.

In mode 2, knowledge acquisition is regarded more as a mutual learning process between the various stakeholders in the research. The aim is not just to propose scientific claims and models pertaining to natural and human phenomena as well as the interaction between them, but also to say something about how this knowledge is interpreted by social actors and how it can be implemented within society. The empirical cycle is then combined with hermeneutic and/or action-oriented research and/or design cycles in which the stakeholders try to develop strategies to bring the theoretical knowledge successfully into practice.

In our efforts to find solutions, both types of research have a valuable role to play. But we also need to bridge the traditional gap between them and combine them. One obvious starting point is to leave behind *dualism* as a fundamentally unfruitful basis for a scientific knowledge theory (Shackley et al., 1996: 203-205; Fay 1996: 223-246). In the previous chapters, we have shown how quite a few dichotomies that are often taken for granted are problematic, as they turned out to be more apparent than real. The subject-object dichotomy between observer and observed is one example of a dualism that we can discard as untenable. The same can be said of similar dichotomies such as objective structures versus subjective actions, the natural versus the social, knowing versus acting, theory versus practice, expert knowledge versus lay knowledge, facts versus values. Particularly within the context of studies that transcend the traditional boundaries between the social and natural domain, it is counterproductive to hold onto the either/or thinking underlying such terminology. Below, we introduce an approach that surpasses this outdated dualistic thinking and instead takes as its point of departure a more holistically focused, integrative complexity thinking.

5.1.1 Methodological Implications of Complexity Thinking

The challenge we face in research projects concerning 'wicked' problems is that they entail the study of open systems in which many social, technical and natural processes co-exist, co-evolve and have an impact on each other in overlapping time scales and levels of organisation. They involve discontinuous, qualitative change as well as cascade effects whereby change strongly and rapidly feeds back into the conditions for further change. More often than not, these problems unfold over decades or more, so there is no relevant 'short run' for a model to operate in. Nor is there the possibility of cutting the system into distinct levels of organisation that can be studied separately. As almost everything is changing with everything else, the effects of our interventions can only be studied against the background of an ever-changing external environment (Andersson et al. 2014: 152).

A word of caution might therefore be in place here: the methodological implementation of complexity thinking is not, and should not, be regarded as a universal remedy to our complex problems. Characteristics such as self-organisation and emergent system properties make micro-level explanations of macro-level system behaviour impossible and render ultimate solutions unattainable. A complexity-based integrative approach fundamentally rejects the linear process as a basis for research design and practice and requires the acceptance of the impossibility of knowing all relevant facts about evolving systems (Morin 2008; Cartledge et al. 2009). At best, we can carry the hope of finding temporary solutions (Osberg et al. 2015: 215). What we need, therefore, is a dynamic framework within which current research practices can be questioned and reconstructed and within which the wide range of conceptual and practical ideas relevant to complexity-based integrative research can be implemented.

Methodological innovations are needed, for instance with regard to the design and interpretation of real-world experiments, the validation of the complexity of knowledge, and how to better account for the diversity of values in evaluation methods (Hirsch Hadorn et al. 2017: 449). It is impossible to provide a blueprint, but a few guiding principles for the design and implementation of complexity-based integrative research projects can be given. These include:

- a systems approach;
- thorough problem framing and boundary setting;
- linking abstract knowledge to case-specific knowledge;
- recognition of the gaps in our knowledge;
- a sophisticated understanding of the uncertainties related to our knowledge;
- a conscientious attention to values;
- a focus on action, implementation and transformational change; and
- an understanding of the nature of collaborations (Hirsch Hadorn et al. 2006, 2017: 431; Pohl & Hirsch Hadorn 2007; Van Kerkhoff 2014).

Conventional research approaches usually seek to abstract from real-world complexity and create idealised conditions in order to try to control as many of the variables as possible. This type of research is certainly of value, particularly with regard to science's aim to produce some generalisable knowledge. But it does not suffice if we want to bring forth solutions to our concrete problems. To enhance the actual implementation of the available knowledge and solutions, it is best to position the researcher as a participant within the real-life complexity. For it is only when research is placed inside the complex system under study that researchers can actually engage and deal with the complexity of the issues of concern rather than merely observe it (ibid.: 144-153, Pohl & Hirsch Hadorn 2007: 112).

5.1.2 Engaging in Complexity

Most of the literature, especially the academic literature, is about what Morin (2008) calls 'intellectual complexity' and much less about 'lived complexity'. However, the plea to put science in the service of society is encouraging action-focused approaches that require researchers and their stakeholder partners to 'live' complexity as a new paradigm for decision-making in communities and institutions (Rogers et al. 2013).

Traditionally, we try to solve problems by making use of the available knowledge within the scientific disciplines. Various experts study the issue and demarcate and research the problem from their own disciplinary perspective. Their findings and proposed solutions are assessed using the conventional criteria of objectivity and value freedom. Besides this group of specialists there are the lay people, who are usually not experts on the topic under study but who nevertheless have an opinion about the issue. More often than not, they express their concerns about the way the research is executed or about the proposed solutions. Since this group has less specialist knowledge, it is considered to be dependent on the experts. The scientific experts are supposed to take the lead and the other stakeholders are expected to follow – they are informed but are usually not actively involved in the research process.

Considering the societal importance of so-called 'wicked' problems, it is important to think of adequate ways to engage the non-academic stakeholders in our search for solutions. This is what proponents of action research, and particularly participatory action research, have advocated for several decades now. It is also the stance taken by transdisciplinary researchers, whose approach to a certain degree resembles that of action researchers. If there is a difference we could point out between modern-day transdisciplinary research and action research, it would be that besides systems thinking, complexity thinking is usually incorporated in the former, with occasionally design thinking as well, whereas in action research this is not customary.

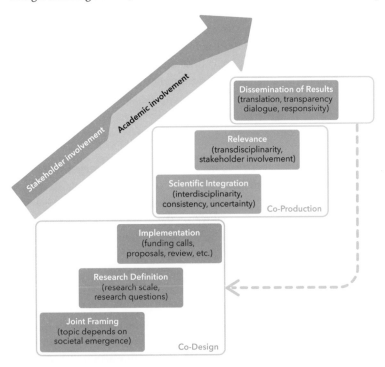

Figure 5.1   Co-design, co-production, co-creation within transdisciplinary research
Source: Mauser et al. 2013: 427

The point that transdisciplinary and action researchers are making is that we need to find ways to involve non-academic stakeholders in co-designing research agendas and co-creating scientific knowledge and to collaborate together in the implementation of that knowledge within society (see figure 5.1). For one thing, non-academic stakeholders often have invaluable contextual knowledge that has a distinctive added value to that of the academics. Another reason is that, in the efforts to tackle the grand global challenges, the interests of the broader group of stakeholders need to be taken into account at least as much as those of scientific researchers. Communities need well-founded, robust information to support their claims and to help them become empowered to strive for sustainability and enhanced well-being (Wilmsen 2008).

A co-creation process generally starts with co-construction of the research object and collaborative problem framing (see figure 5.2). As a systems-based approach, complexity thinking emphasises the importance of considering performance in terms of the whole system rather than its component parts. Nevertheless, a key task is to define and demarcate the system under investigation and carefully consider the gains and losses as well as inclusions and exclusions that are made in this boundary-setting work. Working with stakeholders to collaboratively define the boundaries of the system is an important starting point. Various affected groups will often frame a problem differently, reflecting their specific interests, context and experience. Boundaries are at once practical (in terms of resources available to conduct the research), strategic (in terms of whose interests can or must be met) and intellectual (in terms of which questions are feasible to address, given what we know) (Hirsch Hadorn et al. 2006: 125, 2008; Van Kerkhoff 2014: 146-148).

The scientific researchers work together with other stakeholders to find explanations for the pressing issues and to develop suitable solutions. Instead of turning a blind eye to the various interests and the dominating power relations within the research process, these are explicitly made topics of discussion. The scientists' main interest will be theory or model construction; they want to be able to produce publications based on the research findings. The other stakeholders are probably more interested in finding practical solutions to their concrete problems. The challenge is to do justice to these various interests without jeopardising the process of knowledge acquisition.

What is needed here are research programmes that formulate explicit goals with regard to the dissemination, adjustment, translation and implementation of knowledge from academia to society. This kind of research seeks to both gain theoretical insight and engage in normative reflection, and offers the scientific evidence that is needed to develop effective action strategies for our actual problems. This means that general 'evidence-based' scientific knowledge and argumentations are combined with insights based on especially interesting examples – the 'best practices'. To be able to do this, actual issues must be reviewed in their specific time and context, and the informal though possibly valid argumentations and everyday experiences of the involved stakeholders within those specific contexts should be seriously taken into consideration.

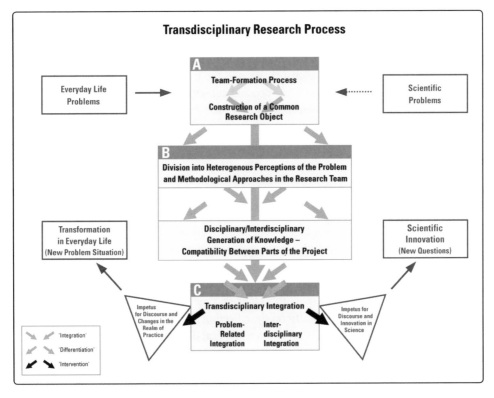

Figure 5.2   The transdisciplinary research process
Source: Bergmann et al. 2005: 19

In contrast to the still-dominant research approach that is characterised by fragmentation between disciplines, between researchers and practitioners, and between research-based knowledge and research-oriented action, the complexity-based research approach presented here offers a foundation on which we can design more integrative research projects (Van Kerkhoff 2014: 153). Researchers that are embedded within the systems they are examining must be prepared to react and respond to emergent findings in nonlinear ways (ibid.: 145-146). This means that the linear process is replaced by a recursive model in which the theory and methods are repeatedly tested by applying them to practice. When necessary, the underlying assumptions are modified and adjusted to try and make them more adequate (Pohl & Hirsch Hadorn 2007: 116). All the while, the scientific researchers will be negotiating with experiential experts and other stakeholders to judge what is the best way forward.

5.1.3   Varying Regimes of Justification

No single expert or discipline is able to provide an overview of any of the 'wicked' problems we face. The challenges are so huge and all-encompassing that it is impossible to separate 'pure' factual aspects from values, morals and ethical aspects. Moreover, the search for solutions to the problems also involves the question of the legitimacy of the solution (is it a good solution?). Part of this is a matter of fact (does the proposed intervention indeed offer a solution to the problem, given the cause and effect relations and the feedback loops and so forth?). Yet there is also a normative aspect (might the solution perhaps lead to any unintended, unwelcome side effects? Will it perhaps result in an unequal division of the available resources?). Justification is not only about 'what is the matter' (which relates to facts), it is also about what is 'good' and what is 'bad' when searching for a solution to a particular problem (which relates to values and moral judgement).

It is not legitimate, therefore, to exclude 'lay' people from the process; they have just as much of a right to decide what is morally good or ethically bad. Scientists have no exclusive expertise when it comes to moral judgments. In public debates and transdisciplinary research projects, both experts and lay people do their best to interpret and evaluate issues. This diversity of input makes it even harder to find solutions to 'wicked' problems.

Gerard Verschoor (2009: 146-155) presents various frameworks for arguments that interested parties present in debates about complex controversies. He calls these *regimes of justification*: value-charged frameworks in which arguments are proposed to support a particular vision or position regarding a complex issue in a public debate. What debaters invariably try to do is to generalise the value that they attach to a particular aspect of the question (or the solution). They do this by appealing to others to support their generalised argument for the common good. Stakeholders resort to arguments like these to persuade a wider group why something is good or bad or right or wrong for all concerned. They have to try and convince others that their own personal vision is relevant and legitimate for the general welfare.

There are various value frameworks or regimes of justification, each presenting their own vision of what should be considered a core value in assessing what are adequate and good solutions. They include:

- *Justifications based on industrial value*
  In this regime, matters are viewed as positive if they contribute to society through efficient production. The application of expertise, technology, standards and methods of limiting risks are highly valued in this regime. Adherents of this regime put their trust in science and are less interested in uncontrollable, private and local solutions.
- *Justifications based on market thinking*
  This regime is characterised by neoliberal ideas. Competition and market forces are viewed as preconditions for economic welfare.
- *Justifications based on equality of citizens and solidarity*
  Proponents of this regime oppose free market arguments and the neoliberal system. Instead, the importance of equal access to markets and collective welfare is emphasised.

- *Justifications based on traditions and the local community*
  In this regime, particular value is attached to local connections, conventions and customs; objections are raised to regulations imposed from a non-local level. New technologies are often viewed as a threat rather than a potential benefit.
- *Justifications based on 'green thinking' and pro-environmental views*
  Proponents of this regime attach high value to paying respect to nature, support sustainability, and try to preserve bio-diversity and a healthy environment for future generations. All perceived threats to these are considered negative.

These value frameworks, or regimes of justification, enable us to make a distinction between various types of arguments that are brought forward in a debate. They help clarify why it can be difficult to find a solution to a complex issue.

Various Groups of Stakeholders Employ Different Regimes of Justification

Scientists involved with GMOs often use a range of knowledge claims, as the controversy regarding genetically modified corn in Mexico clearly demonstrates. Verschoor (2009) shows the role played by regimes of justification here and how adherents of the various regimes justify their views.

Proponents of modified corn emphasise efficiency and productivity and justify their position by arguing that this will enable us to feed the growing global population. This is industrial regime and market regime thinking, which is based on goal achievement and profitable investment. These regimes of justification are employed by scientists convinced that the target of efficient production is attainable and who insist that GMOs are safe. Building upon evidence-based scientific research, they make universal claims on the basis of which they try to attain general certification procedures and international treaties.

Opponents of modified corn propose quite different, often contradictory arguments. Scientists working for NGOs such as Greenpeace point to the danger of creating a monoculture, which is a danger to ecosystems and bio-diversity in Mexico. Employing the green regime, they advocate adopting a careful attitude regarding nature and campaign for sustainable food production. In their view, we should use agricultural methods that have a minimal impact on the environment, which rules out genetic modification.

Criticism of GMOs is also voiced from the regime whose watchwords are civil equality and solidarity. These groups criticise the way that major multinationals disadvantage small-scale farmers by making it impossible for them to compete. Many multinationals take out exclusive patents on special kinds of corn seed, forcing farmers to buy from them and so risk losing their independence. Proponents of solidarity present the consensus regarding the value of equality and collective well-being as legitimation of their view that GMOs should not be permitted.

And then there is the regime that presents local values and traditions against the main arguments of other regimes. What is at stake here is the protection of certain cultures and respect for those who are living in harmony with nature. Modern agriculture, with its industrialisation, economies of scale and globalisation are considered as threats to these basic principles. And bio-technology is seen as an instrument wielded by big business, focused purely on market domination and the subduing of nature.

While the GMO conflict is often presented as a debate between those for and those against science, this is a misleading portrayal of the actual situation (Melchett 2012: 1). In the industrial and market regimes, the argumentation is mostly related to scientific evidence, although certain underlying values will definitely influence their way of reasoning (e.g. the idea that technological solutions and increasing production are the best way to solve the food problem). In the regime based on the core values of civil equality and solidarity and the regime in which local values and traditions are prioritised, the emphasis is on the normative aspect. Yet in these regimes, scientific arguments will also be used to justify their claims. In the same vein, both scientific and normative arguments are advanced in the green regime (e.g. by pointing to disappointing numbers with regard to anticipated GMO yields or the increase in suicides under farmers working with patented seeds). So although the proportions may differ within each justification regime, in most cases there is a combination of scientifically supported arguments and normative arguments.

Even within one group of stakeholders – e.g. scientists – we see that widely varying regimes of justifications are used. Scientists working within industry often use completely different justifications than those working within the environmental movement. Obviously, both types of scientists differ strongly in how they perceive reality and what they value as important. But if all knowledge is dependent upon certain values and presuppositions, how can we be sure that the path we are taking is leading us in the right direction? When we have two plausible explanatory models, or when diverging approaches point to different outcomes, how do we decide which is the right one, or at least which is the best? Can we compare them, or are the insights and solutions brought forward by one group of scientists in essence incomprehensible for or unacceptable to another? These are the questions that we will address in the next section.

5.2   Science in Progress

The conventional image of progress is that scientific knowledge grows cumulatively through continuing research. Thus, every new insight broadens our body of knowledge to add to an ever-growing foundation of facts. While refutations of our hypotheses may force us to conclude that out theory is on the wrong track, falsification usually also results in new ideas about the actual situation. So our supply of knowledge steadily increases, and little is lost. The confirmation level – the degree to which developed theories are confirmed – gradually rises. And the idea is that theories with a high degree of *corroboration* – i.e. higher than the level of falsification of the hypothesis or theory – are essentially better (Popper 1963, 1972).

Similarly, a representational model is held to be true if it does not lead to false conclusions about the target. A model is validated if we are convinced that there is an appropriate fit between the dynamics of the model and the dynamics of its target system (Winsberg 2010: 105). And it is thought to be empirically adequate if it is complete with respect to all the observable or measurable aspects of the target (Suárez 2004: 776).

However, as shown in chapter 2, this image of scientific progress is too simple. The correspondence of truth and its accompanying idea of verisimilitude is problematic, and we cannot rely one hundred percent on the representational function of our models either. This leaves us with the question of how we can establish whether the chosen approach leads us (closer) to the truth – or, in the computational variant, how we can ascertain whether our successive projections and model simulations are improving. We know that the coherence theory of truth helps us in this respect by emphasising the reliance on consistency in our explanations and interpretations. But it does not offer a watertight solution either. Apparently, we have to learn to live with fallible knowledge and temporary solutions. Yet we would still like to be able to determine whether indeed there is progress instead of merely continuous development. In a time of increasing complexity, what can we do to ensure that we achieve robust explanations and solutions that will stand the test of time and changing circumstances? This is the challenge we will now investigate further.

5.2.1   How to Determine the Reliability of Knowledge

When scientists argue that their simulative models are trustworthy, they do so on the basis of the conviction that these models agree with an actual, real-world system or at least optimally depict its behaviour. Thus they are indirectly arguing that the models approximately reveal the content of the best theory of that real-world system (Winsberg 2010: 25). However, the deciding factor is not which approach is most true to theory but which approach provides the most reliable information about the real-world target and produces the best solution set as the outcome of the simulation. It is not difficult to recognise the instrumentalist position here: the solution set considered to be the best is the one that best reveals the features of the system that are important for understanding it (ibid.: 9). Thus, according to Eric Winsberg, reliability is an alternative concept to truth (ibid.: 132).

Establishing the reliability of a model does not depend on just one method. Rather, it is the simultaneous confluence of all efforts to show the reliability of the model. By showing how it agrees with the governing theory and how it fits into the web of the available empirical findings and previously accepted data, how it matches the results of the prediction, how it conforms to mathematical rules as well as common sense intuition, we gain confidence in the results. In this sense, model-building techniques such as simulations are self-vindicating (ibid.: 45, 122).

Although Winsberg explicitly talks about the reliability of models, his proposal can be read to pertain to the whole gamut of knowledge that can be gathered with empirical, hermeneutic, modelling, action and design cycles. It portrays a way of looking at the validation process that resembles the proposal to work with a network model of correspondence and coherence as the basis for our accepted body of knowledge (see section 3.2.4) – a network model that may be fallible but still strong enough to support us in our claims that we hold robust knowledge.

As a reminder of the need for humility, Batty and Xie (1997: 191) warn us that it is unlikely that today's models will generate acceptable levels of performance suited to existing patterns in the same way that traditional, less complex models can. Models always distort the target system in one way or another, and inevitably there are aspects of uncertainty and ambiguity attached to the relation between model and target. These complexity-inherent characteristics lead Mauricio Suárez (2004: 776) to claim that models may be non-isomorphic and dissimilar compared to the target system and still license true conclusions. That is because particular mechanisms tend to dominate the system dynamics at given points of time, which may cause certain parameters to disproportionally affect the system and thus show odd patterns of behaviour that do not fit with the model's performance. We can compare this to the problem encountered in chapter 2; just as we do not know whether to blame the theory or an auxiliary hypothesis when a prediction fails to match the observed data, we do not know whether to blame the underlying model or to blame the modelling assumptions (Winsberg 2010: 24, 105-106).

Since many models incorporate random elements, the results of just one run cannot be relied on. By running several simulations, we can take some steps towards determining the relative importance of a given mechanism for the system dynamics. It is necessary to establish that the results are robust with respect to different random values. Nevertheless, if the model describes the target well enough, then unexpected or undesired conceptual patterns from the target should not make the model unreliable or fragile. If the target does complex things, and the model can account for this or does not fail to fulfil its claims, then it can be considered to be robust (Pieters 2010: 172).

However, in view of the constant transformation within complex systems, we must be satisfied with a weaker type of knowledge claims. As Winsberg (2010: 27-28) says: 'When it comes to complex systems, we simply cannot bend our theories or cognitive will – they will not yield results with any mechanical turn of a crank.' In complex systems, predictability is limited, so there will always be unexpected outcomes from any intervention or change. Therefore, Batty and Xie (1997: 191) hold that the emphasis in applications should not be on fit but on feasibility and plausibility. Having multiple strategies, plans or responses in place allows for more flexibility in the system to react and adapt to emerging or unpredicted outcomes. The uncertainty and unpredictability of change in complex environments is not regarded here as something to be reduced or minimised; it is simply a dimension of research that we must take into account. Researchers into complex issues should expect the unexpected and should be prepared to deal with it (Van Kerkhoff 2014: 148-149).

Paul Cilliers (1998, 2001) accordingly suggests a significantly less universal conception of scientific knowledge: as contextual, local and specific in time and space. Again, the fact that the possibilities for prediction and description are limited does not mean that 'anything goes' and in no way downplays the importance of scientific work. In fact, quite the opposite is true: the fact that our knowledge of a system is only local and temporary emphasises the importance of knowing how to learn about a system (Byrne & Callaghan 2013). The fact remains that the world can be known, even if that knowledge is contextual and time limited.

A Global Food Approach or Local, Context-Bound Solutions?

In the 1970s, considerable effort was made to generate a 'Third Agricultural Revolution' or 'Green Revolution' to increase food availability worldwide by producing higher yields, more calories and more proteins. The failure to realise such a revolution in the developing world has led some scientists to conclude that the formulation of the food issue as a global problem proved a bridge too far (Maat 2011: 173; Fresco 2009: 379). Recently, there have been calls for more regional, context-specific approaches. Consequently, opinions regarding agronomy and agricultural and food technology have changed.

Today, agronomy is the science of agriculture, ecosystems and the environment. Technology is no longer seen as an external source of change but as a fine-tuning of ecological and technical processes in a given socio-economic and cultural setting. More attention is given to specific conditions in a particular country such as the type of food systems, the degree of technological permeation, the proportion of the labour force in agriculture and other relevant variables (e.g. cultural or gender issues) so that specific, appropriate strategies can be applied. It is acknowledged that, to accommodate global changes, agronomical factors need to be coupled with nutrition, economics and social science, involving a variety of aspects such as food production and processing, access and utilisation. Only when we do justice to the specific contextual conditions without losing sight of the intricate linkages between food systems in the world can we hope to develop strategies that are simultaneously nationally and regionally based and internationally inspired (ibid.: 383-384).

Paradigms, the overarching theoretical frameworks that reflect the zeitgeist of an era, steer the work of scientists in a certain direction. Kuhn (1962) considers a paradigm to be decisive in the sense that it determines our outlook on the world. The standards and values used to determine which paradigm is the best can vary significantly across different paradigms. When one paradigm is replaced by another, a fundamental shift takes place in the way we think about and view reality (Kuhn speaks of a 'Gestalt shift'), so that our whole worldview changes.

Given the enormous differences in underlying visions, it is in Kuhn's view practically impossible to compare theories from different paradigmatic traditions: they are *incommensurable*. Facts that form convincing confirmations or refutations of a particular theory viewed from one paradigm may be quite the opposite in another paradigm. Kuhn therefore argues that we should not expect too much from paradigm shifts. Since they are inherently linked to the underlying system of assumptions, paradigms cannot be compared and weighed independently, as each paradigm has its own set of assumptions.

In effect, Kuhn questions the very idea of the cumulative growth of science. After all, if we can no longer state for certain that one system is better because it offers more true theories and explanations than another, we must necessarily abandon the idea that successive paradigms bring us closer to the truth. Apparently, there is no neutral language or independent standard by which we can measure the different paradigms. It seems that we can only state something meaningful about the social and communicative processes that govern the successive scientific paradigms (Kuhn 1970: 20). If that is so, then the idea that we can develop scientific knowledge in an internally regulated rational learning process suddenly appears rather vulnerable.

Kuhn always tried to downplay this implication by pointing out that some scientific values are shared by everyone, whatever the paradigmatic tradition in which they work. These are values such as theoretical simplicity, predictability and the demand for consistency (cf. the coherence theory of truth). Popper also vigorously protested against attacks on the portrayal of science as an internally regulated rational learning process. Such attacks obviously challenged science's particular social and cultural status. He called the notion that paradigms form such diverse worlds of ideas that it is logically impossible to translate concepts from one paradigm to another 'the myth of the framework' (Popper, 1970, p.: 56). Popper felt that it was certainly possible to discuss and compare different frameworks logically and critically, and so to determine whether a paradigm shift is progressive and rational.

Modern Synthesis – Proof of the Commensurability of Paradigms or not?

One example of how various contradictory theories are brought together is Modern Synthesis, also known as New Synthesis. Biologists generally accept the theory that knowledge from different branches of biology – such as palaeontology, population genetics, heredity and evolution theory – all share a certain common foundation. This theory is based on Huxley's *Evolution: The Modern Synthesis* (1942), which still remains one of the key reference books in biology.

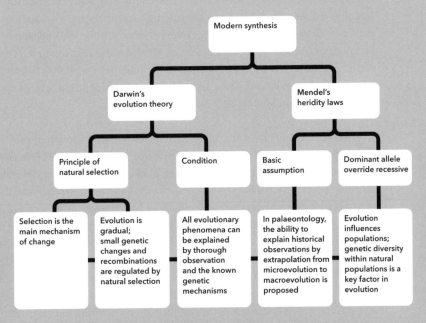

Figure 5.3 Modern Synthesis as overarching paradigm in biology

Modern Synthesis bridged the gap between the work of experimental geneticists on the one hand and naturalists and paleontologists on the other. It states that:

- All evolutionary phenomena can be explained in a way consistent with known genetic mechanisms and the observational evidence of naturalists.
- Evolution is gradual: small genetic changes regulated by natural selection accumulate over long periods. Discontinuities among species (or other taxa) are explained as originating gradually through geographical separation and extinction, not by sudden changes from one generation to the next.

- Natural selection is by far the main mechanism of change; even slight advantages are important when continued. The object of selection is the phenotype in its surrounding environment. The role of genetic drift (i.e. drifting away from original characteristics by a random spread of genes through sexual and asexual reproduction) is equivocal.
- Thinking in terms of populations rather than individuals is primary: the genetic diversity existing in natural populations is a key factor in evolution. The strength of natural selection in the wild is greater than previously expected; the effect of ecological factors such as niche occupation and the significance of barriers to gene flow are all important.
- In palaeontology, the ability to explain historical observations by extrapolation from microevolution to macroevolution is proposed. Historical contingency means explanations at different levels may exist. Gradualism does not mean a constant rate of change.
(http://en.wikipedia.org/wiki/Modern_synthesis)

The fact that the previous contradictions have been resolved in Modern Synthesis may also serve as evidence that there was no fundamental conflict between the different components. If this point of view is adopted, then Modern Synthesis is not to be regarded as a succession of paradigms but rather as an integration of various theoretical frameworks into a coherent whole.

But it should be pointed out that the Modern Synthesis did not resolve all the contradictions. The notion that different types form only *after* different types have separated reproductively is still a matter of considerable debate. This may support Popper's argument that fundamentally different frameworks – i.e. paradigms – can still be discussed and compared in a perfectly logic and critical way.

All in all, evolutionary biology has undergone quite a transformation over the past few decades. The new theoretical landscape that has emerged has led some to claim that a shift is taking place in the direction of a so-called Extended Synthesis. It has invigorated a class of biological theories that maintain a grounding in quantitative theorising but that also has close similarities with qualitative theories in the social sciences (Pigliucci & Müller 2010). Whether Extended Synthesis must be seen as merely an extension of the reigning Modern Synthesis paradigm or is at least a turning point, or perhaps even a tipping point, remains to be seen.

Popper objected to Kuhn's suggestion that the rationality of science assumed the acceptance of a shared framework. He did not believe that rationality depends on elements such as a shared language and shared assumptions, nor did he believe that rational discussion and rational criticism are only possible when there is agreement on the fundamental principles. He paid little regard to the idea that scientists have to at least agree that the critical rationalist research approach represents the research principles that are constitutive of science. Yet we have seen that this is not the case. Science is neither uniform nor universal, and there is no single scientific method. There are diverse views on science, all of which lead to various approaches in which different rules and procedures apply. In other words, there is pluriformity in science. We have seen how difficult it can be to determine whether it is possible to reconcile overarching theoretical systems that belong to the same paradigm (like evolution theory and heredity, see the example in the box). By extension, it must be even harder to compare approaches from different paradigms. Therefore, it will come as no surprise that commensurability to this day remains a point of discussion in philosophy of science.

With the introduction of computational models, this discussion has been reinvigorated. Models are comparable to theories in the sense that they are dynamic tools offering explanations and predictions of the target system that are up to the community for critique. A researcher or collaborating group of researchers within this community makes decisions about what phenomena are interesting or confusing, what questions are worthy of pursuit, what modelling decisions are acceptable and appropriate, and what kind of knowledge will fulfil the researchers' goals (Giere 2006, 2010). Often, models are partial renderings of a system under investigation, and usually many different kinds of models are needed to help map out a single system. In such cases, it is not always possible to add corrections to a stable structure to increase the accuracy of the model. The addition of parameters would result in a new model that presents a radically different account of the system and its behaviour. Hence, in describing the system's processes, we can end up with a number of models that are inconsistent with each other (Morrison & Morgan 1999: 28) or even a model construction that entails inconsistencies (Winsberg 2010: 86). With no ultimate criterion to assess their validity, we are left here with the same dilemma as back in the old days: how to establish which model (or configuration of models) is the best?

Peter Godfrey-Smith (2006: 739) manages to point to an advantage in the current situation, though: with much day-to-day discussion focusing on model systems, disagreement about the nature of a target system is less liable to impede communication. The model acts as a 'buffer', enabling communication and cooperation between scientists who have different commitments with regard to the target system. So some situations that might potentially generate linguistic incommensurability are prevented from doing so because the model system provides common ground. In this sense, models might even function as devices that can help overcome potential communication problems in science.

The flip side of this solution to incommensurability problems, however, might be that it leads to a kind of inertia in some scientific fields. This can happen when a group of scientists makes full use of its ability to hold onto a model while the surrounding knowledge changes. Features of the background context that initially made the model a strong contender may be lost or abandoned, but the model is nonetheless retained. Maybe some repairs are performed to redefine the relation of the model to the target system in order to be able to cling on to the model. In the case of modelling, this seems easier and even more attractive than in the case of theories. However, this adjusted retained model would never have been the result of work that started afresh from a new set of background beliefs. So the 'buffering' seen in model-based science, stemming from the flexibility in its design and construction, may have both good and bad features. Whereas it enables continuity, makes communication easier, and prevents scientists from having to design completely new models each time the findings do not fit with the model, it sometimes offers a solution that is merely apparent but not real (ibid.).

5.2.3 Progressive Research Programmes and Problem Agendas

Another effort to tackle the incommensurability problem that is worthwhile discussing here is Imre Lakatos' *Methodology of Scientific Research Programmes* (1978). He tried to sketch how scientific progress can be guaranteed by demanding that paradigms take on the form of subsequent progressive research programmes. A research programme is considered to be progressive when its 'hard core' entails theories with more explanatory power than its predecessor. It remains progressive as long as it generates strong claims and impressive accurate predictions. It becomes regressive when it is only focused on retaining its protective belt of auxiliary hypotheses in order to prevent falsification of the overall programme.

> ### *Problem Agendas in Developmental Biology*
>
> An approach that seems similar to Lakatos' proposal with regard to progressive research programmes is that of Alan C. Love's problem agendas (2014). Love claims that developmental biology owes its stability and structure less to the presence of theories than to the scientific questions that drive the research process. He calls this the erotetic organisation of developmental biology (erotetic = pertaining to a question). Love posits that problem domains with problem agendas determine the course of research over an extended period and give structure to the knowledge acquisition process.
>
> Critics such as Thomas Pradeu (2014) point out that science is not just about problems; in fact, it is more about the answers that are found. Love does not deny that theoretical knowledge is employed in developmental biology. In fact, a whole range of theories is cited to find explanations and to come up with predictions. But Love's point is that these theories are not what drives

research in a particular direction or what gives structure to the acquired knowledge. The direction is determined by the questions derived from problem agendas. These problem agendas provide a guide to both research practice and theory formation, and they help to coordinate ideas generated from different biological terrains. They have a supportive function in cross-disciplinary research. Problem agendas can enhance the process of knowledge integration because they have common criteria regarding the nature of adequate frameworks (Brigandt 2016: 611-615).

Ingo Brigandt (ibid.) thinks it is unlikely that we will find one uniform development theory. Organisms are too complex and the development of the different taxa too varied. The diversity of the theoretical approaches in developmental biology is therefore broad. He argues that different theories, each pertaining only to a certain sub-domain, will continue to exist.

Therefore, Brigandt thinks it is more interesting and fruitful to shift our focus from the empirical content of theories to ideas that have an impact on the theory behind the methodology. Instead of looking for representations of phenomena that occur among organisms, more attention should be paid to generating explanatory and predictive claims, to the formulation of criteria that clarify at what level explanations and predictions are adequate, and to modelling strategies. All this would be undertaken in the hope that such an approach will eventually lead to the emergence of a theory that would enable us to organise developmental biology as a whole, as a discipline.

The development of science, often referred to as scientific growth, is supposed to be an autonomous process driven by rational criteria. In the ideal situation, the rationality of science rests on the fact that each step in theory formation is based purely on logical argument or empirical facts. Yet if it is about finding a reason for calling a research model or programme progressive, philosophy of science cannot rely exclusively on such priorly given principles. As Lakatos points out, there is no such thing as 'instant rationality'; what is regarded as scientific rationality is determined by the academic community. Internal criteria – i.e. rules resulting from the scientific research practice that are brought forward from within the scientific community itself – are insufficient to determine the value of a particular research model or programme. In addition to the internal rules we also need external rules, i.e. rules derived from the concrete contexts and practices to which the scientific explanations pertain and in which the generated knowledge is used. Both internal and external factors will have to be taken into account in the assessment of the value of a particular research model or programme.

This take on scientific rationality does not fit well with the traditional model of science supported by Popper and to a certain extent also by Kuhn. However, in an integrative, complexity-based science built on a broad conception of rationality, the standpoint that the value of a research model or programme is established by reference to both internal and external criteria is perfectly defensible. It means that the interests and values that societal stakeholders attach to research are taken just as seriously as the interests and values that scientists themselves are committed to when doing research.

5.3 Quality Criteria for Research into Complex Issues

Evidently, the enormous complexity of urgent problems makes it difficult to determine whether our explanations and analyses are conclusive, let alone whether our solutions are final or even appropriate. Moreover, in transdisciplinary research, the community assessing the validity and value of scientific knowledge is broader than that involved in monodisciplinary, multidisciplinary or interdisciplinary research. Besides the scientific community, there are also other stakeholders assessing the quality of the knowledge gained during the research project. For the sake of simplicity, we have hardly considered non-scientific stakeholders so far. However, complex problems such as the global food issue involve various groups with diverse interests, each trying to show that their view on the problem is the most legitimate one (see regimes of justification in section 5.1.3). We know now that they employ scientific facts in their arguments as well as appeals to universal norms and values that many share. Hence, in today's society and scientific world, the logic of ultimate solutions no longer works. The criterion of trying to find an ultimate solution has been replaced by alternative criteria, such as the feasibility of solutions, the level of consensus attainable for different options, and the contribution of the proposed solutions to the overall sustainability of a system (or sub-system). Below, we investigate the methodological implications of this shift and what this means for the quality criteria with which scientific knowledge is measured.

5.3.1 Objectivity Defined as Critical Intersubjectivity

The realisation that reality and our knowledge of reality are inextricably intertwined has led to an epistemological theory that combines perspectivism with fallibilism. Accepting that research results can never fully represent or cover reality, the idea that research can be objective is abandoned. Instead, objectivity is reinterpreted and redefined as 'critical intersubjectivity'. This intersubjectivity can be achieved through a willingness to be held accountable for the perspective and the position taken as a scientific researcher and an open and honest attitude towards constructive criticism (Fay 1996: 212-215; Brown et al. 2010: 42).

Viewing research objectivity in this way, it actually consists of a social process of continuous criticism. This process is intersubjective because it is a constant dialogue between researchers. Scientists review the theories and research findings presented by their colleagues. Likewise, they need to be prepared to review their own theories and research findings when straightforward criticism warrants adjustment. So criticism is inherent to the process; developing knowledge is constantly being examined and systematically tested to see whether a particular explanation, simulation or method still pertains. Objectivity is therefore an aspect of cooperative discussions focused on the collective research into the value of different theories, models and approaches from a thorough, open research attitude. The precondition for such an attitude is that researchers are able to distance themselves sufficiently from their own subjective perspectives and to take into account other perspectives.

Fallibilism has taught us that research results can never be indisputable. So we have to recognise that we can only make qualified validity claims and that our models are not permanently reliable. Future developments may make today's theories and models obsolete. This is why the process of critical intersubjective reflection is continuous; it is a never-ending practice that is fundamental to appropriate scientific research.

Viewing objectivity through this lens represents a shift from substantive to procedural adequacy. Objectivity no longer relates to the results of research but is thought to be dependent on the process, on the question of whether the research is adequate. It is the method of scientific analysis, not the conclusions, that can be regarded as objective or not.

Quality assurance of knowledge produced in transdisciplinary research goes even a step further. The point of departure here is that our social and scientific knowledge and practices are not only produced by a multitude of interlinked relations but are also constructed alongside each other and support and strengthen one another. Scientific perceptions, paradigms and methods emerge from mutually constitutive processes, whether with other scientific fields, policy institutions, user groups, industry, public agencies, the media and so on, rather than being solely the result of internal scientific advances. An implication of this approach is that the characteristics of scientific knowledge (for example, levels of aggregation, defined system boundaries, and representations of uncertainty) embody unstated assumptions about the needs, capabilities and structure of society or, more particularly, the policy world. This being the case, it only seems fair to explicitly involve societal institutions and the lay public in the construction of research agendas and problem definitions. And the same can be said for the assessment of the outcomes of the research projects (Shackely et al. 1996: 208, 218).

Quality assurance of knowledge produced in transdisciplinary research must take into consideration the mutual influences between the empirical level of systems knowledge, the purpose level of target knowledge and the pragmatic and normative levels of transformational knowledge (cf. the broad concept of rationality in which facts, norms and values are taken into account). This calls for collaboration between researchers and social groups in order to be able to regularly validate and adapt empirical models in concrete situations. Repeated efforts to build a consensus about purposes are also essential, as are the constant monitoring of the effects and the adaptation of transformation strategies in the implementation phase of the research projects (Krohn & van den Daele 1998).

In a science that has abandoned the old idea of objectivity in exchange for a new definition of objectivity as critical intersubjectivity, researchers must be prepared to be held accountable (Brown et al. 2010: 42; Fay 1996). They have to be willing to justify the positions they adopt in two ways: they must be able to give an account of the approach they use and of the position they take as researchers (ibid.: 216-218).

The first type of accountability – accountability for the approach used in the research – involves more than just explaining one's own epistemological assumptions. It extends beyond that to explaining the wider conceptual assumptions such as the cognitive commitments and frames of reference that influence the way the research is designed and executed (for instance the concepts, materials, models and methods that are used to observe, categorise, characterise, indicate and simulate the world). It also involves the willingness to give an account of what criteria are employed to decide what is important, interesting, productive or significant in the first place. And, after the choice of the research topic is made, the researchers must be able to report on what they regard as evidence, as relevant facts that enable them to find answers.

The second type of accountability – accountability for the position taken in the research – means that scientists explicitly need to think about their relationship to society. A wide range of positions can be taken in this respect, depending on the kind of knowledge acquisition mode the researcher has committed to. Mode 1 involves a quite different view of the relationship between science and society than mode 2.

Socially oriented studies require even greater accountability, as those involved as researched or co-researchers have a right to know the position of the researchers regarding the research topic. The same can be said for the research commissioners involved (e.g. politicians and policymakers) and the wider public. It is important to think about what is the best way to present the research, considering the target groups and the language they use (e.g. is it going to be an academic presentation or a generally accessible story? Will it be told in everyday language or in a foreign language?). Particularly in transdisciplinary research, which often involves non-academic stakeholders, these are important issues that need thorough consideration (see Bergmann et al. 2005).

Critical intersubjectivity and a reflexive attitude also require scientists to take the social positioning of their research into account. All research – even research without stakeholder participation – takes place in a network of social relations comprising researchers, a varied public and the subjects or objects central to the research. Decisions on who is to be given a voice, who is to be recognised as an authority and why, whose worries and needs are to be met, who will have access to the research material – these are all important decisions because they involve major interests and power relations. An accountable attitude towards the social positioning of their research calls for researchers to continuously reflect on these issues and on the implications of the choices made in this regard.

At first sight, this new view of objectivity may not appear to be so different. After all, science is all about critical research which is published in publicly accessible journals, for which authors are required to conform to well-known scientific criteria such as controllability, transparency and methodological accountability. Their scientific contributions are subsequently subjected to vigorous criticism in anonymous peer reviews. Nevertheless, the redefinition of objectivity as critical intersubjectivity does represent a fundamental change to the critical rationalist paradigm in a number of ways.

First, it represents a new approach in that logic and empirical evidence are no longer the ultimate foundation, while no assumption is made that a completely objective and value-neutral position is possible. This implies, secondly, that the idea that cognitive-instrumental knowledge is the only goal in science must also be abandoned; values and norms inevitably intrude the research process and must be legitimised somehow. As a result, the quality criteria by which scientific knowledge is measured have shifted. The value and validity of knowledge is traditionally dependent on the assessment of the scientific community, but in transdisciplinary research, the assessment and peer review are extended to a wider community of engaged stakeholders (Funtowicz & Ravetz 1994: 578). While logic and empirical evidence remain important guarantees, they are placed in a wider context now.

5.3.3 Searching for Common Ground within Regimes of Justification

As described above, regimes of justification are frameworks that people – researchers as well as other stakeholders involved in complex questions – use to make validity claims and value judgements regarding particular issues. Since a justification is always from a particular perspective, participants in the same controversy may bring forward different arguments stemming from different perspectives.

Regimes of justification may appear to contradict traditional scientific criteria such as objectivity and value-neutrality since they involve subjective, value-charged aspects. Verschoor (2009) notes, however, that no one can be completely objective and value-neutral, not even scientists. Take the neo-classical macroeconomists whose research is based on the notion of the 'homo economicus': a purely rational person who considers no norms or values when making choices and who holds that economic growth is the recipe for progress and a better world. These assumptions are not objective or neutral and may be severely criticised. Examining them from the various value frameworks identified by Verschoor, these assumptions may actually represent a specific justification framework, in this case the market regime. So we could ask: to what extent are the arguments proposed by scientists and based on value-charged assumptions any different from those presented by other stakeholders? In each case, the arguments seem to contain a piece of mode 3 knowledge; and the question of what is the wise thing to do seems to be an implicit part of the underlying assumptions.

What may be considered crucial is the extent to which the arguments can be made acceptable for all, or at least for a large group of people. To that end, the assumptions must first of all be made explicit. An advantage to this is that they then also become transparent and testable. This enhances the quality and the research and argumentation process, which has been shown to be decisive in sorting the wheat from the chaff and separating scientific and well-founded statements from pet notions and unfounded opinions.

By evaluating a research process while relying on various regimes of justification, it is possible to test this against the new scientific validity criteria. It may also be possible to bring together contradictions and apparently irreconcilable arguments in a single coherent justification framework at the meta level. This would involve various experts coming together to perform a value assessment. And where this involves transdisciplinary research, other stakeholders may also be engaged, such as experience experts or local actors directly involved with the issue.

When everyone accepts the justification, it may be treated as an 'objective' and 'value-neutral' generalisation. The perspective has fundamentally changed here. You could say it is the result of the realisation that whether an argument should still be considered subjective if it is maintained by a large group of proponents is as legitimate a question as whether an argument can still be said to be objective if it is criticised by a large number of opponents. Here again, objectivity is defined as: intersubjectively acknowledged as being the case, while in this case, value-neutral means non-partisan, unbiased, and not influenced by subjective interests or for personal gain. Yet the generalisation remains subject to interpretation.

If a successful generalisation can be obtained based on the various regimes of justification, Verschoor's method may be considered an innovative approach. In response to the standpoint that these justification frameworks are impossible to reconcile with the traditional justification criteria of objectivity and value neutrality, we may say that they offer a fruitful way of testing whether the presented knowledge can actually be regarded as objective and value-neutral science.

5.4   Dealing with Complexity

Traditional ways of approaching complex problems by attempting to tame, confine or manage them frequently fail to achieve their goal. This is often due to the intense political pressure accompanying such problems, as the interests involved in the policy decisions related to these issues may be substantial. The factual knowledge about 'wicked' problems is by definition uncertain and controversial. This applies all the more to the values held by the various involved stakeholders, which can be highly diverse.

What should be definitely avoided here is to view the systems involved in complex issues as merely 'objects' – which still often happens even though nowadays they are sometimes regarded as integrated socio-ecological structures. If they are viewed as objects, the subject-object dualism remains unchanged, just like all kinds of related dualisms such as natural versus social, facts versus values, structures versus actions, etc. Consequently, the role of the subject – be it researcher, policymaker or other type of stakeholder – remains unchallenged, which is not a good thing (Shackely et al. 1996: 220).

What Simon Shackely, Brian Wynne and Claire Waterton fear is that if the newly promoted paradigm simply incorporates the same old ambition to manage and control, there is a real risk that determinism will reign in science once more, and this time even more comprehensively than before. Their caution stems from the methodologising tendency they observe in the contemporary uptake of complexity theory. What they see is that whereas a range of methodological tools is developed, a comparable elaboration of institutional development and reflexivity fails to ensue. If this situation continues, complexity thinking may come to imply no more than the substitution of one set of methods, now seen to be inadequate, with another more sophisticated set, without any serious questioning of existing practices and institutions. In that case, the onset of a complexity paradigm gets stuck at the level of a computational turn, and no consequent transformations will take place (ibid.: 201-217). The danger – expressed in chapter 1 – that complexity thinking will become just another case of disciplinary imperialism focused on bringing society into the realm of the natural sciences might then indeed materialise.

The challenging implication of complexity is that we need to refashion our institutions and explore new ways to represent and acknowledge their 'messiness' (ibid.: 203). Now that the old 'certainties' associated with modernist institutions and practices are eroded (Funtowicz & Ravetz), there is ample reason for change. Thus the implications of a real turn towards complexity thinking reach beyond the domain of science itself. What needs to be acknowledged is that complexity resides more in the character of the relations between institutions, experts and a range of publics – including users, customers and those on the margins of inclusive networks – than it does in more material and reified versions of 'reality'. As complexity thinking needs to be incorporated in society's basic design and practice, new forms of institutional mediation deserve to be explored.

Perhaps it could help if we recognise that much of the challenge for institutions lies not in understanding and managing something qualitatively new called 'complexity' but in positively acknowledging their already-existing complexity and 'messiness' and coping mechanisms (ibid.: 221). More ambitiously, policy and user institutions need to be directly encouraged not only to talk about complexity, and then pretend to manage it as before, but also to embody its realisation in developing institutional relations, mediations and identities.

For interdisciplinary researchers of 'wicked' problems, the art is to develop the knowledge, attitude and skills needed to deal with complexity. They need to be able to assess scientific arguments from different disciplines based on their reliability and usefulness. And they must develop the capacity to build bridges between the different conceptual and value frameworks that are used, and mediate between a variety of action strategies proposed by the various parties involved. Since complex problems may not be solved for centuries, if ever, this involves a continuing process of acquiring knowledge, sharing and assessing facts and values, determining positions and building bridges between them, monitoring effects and implementing adaptations. We can try to meet these demands by doing the following:

- use scientific knowledge regarding the issue (knowledge development through mode 1 an mode 2 science) as the basis;
- develop problem-solving processes that cut across various academic fields (multidisciplinarity and interdisciplinarity);
- recognise the principle of systems thinking in which attention is paid to subtle interactions and retroactions (feedback loops) involving different actors and factors (complexity thinking);
- avoid thinking in terms of the dichotomy of systems/structures and action, since the interplay between these is a crucial element (duality of structure, systems thinking, action research, design thinking);
- acknowledge the multitude of perspectives and the various frames of reference and regimes of justification held by different stakeholders groups (transdisciplinarity, mode 3);
- use the variety of available knowledge, knowledge theories and adequate research methods (pluralism);
- reconcile this knowledge in a coherent manner (philosophy of science);
- complement this knowledge with the theoretical, practical and experiential knowledge of other stakeholders (transdisciplinarity, scientific and common sense knowledge, including mode 3 insights);
- generate sustainable future-oriented solutions for the challenges at hand, which are usually characterised by a lack of structure and which depend on emergent system processes where ambiguities and uncertainties invariably apply (systems and complexity thinking; transdisciplinarity, design thinking);
- recognise the normative aspects linked to these problems, such as how to deal with the relationship between humanity and nature (mode 3 questions), as well as how to prioritise proposed solutions (mode 2 questions);
- take into account the fact that it is worth exploring solutions on both the systems side (e.g. maintenance and repair of a given system, the influence of more or less stable structures – referring to the macro-dimension and event causality) and the action side (e.g. changing consciousness and corresponding behaviour patterns of people in the research domain, and the influence of self-organising organisms or human interventions on the system, referring to the micro- and meso-dimension and actor causality).

By combining modes 1, 2 and 3 knowledge acquisition in multidisciplinary, interdisciplinary and transdisciplinary research processes, future-focused researchers can help to find solutions to today's complex issues, however temporary. Where possible, they can also help implement these in the hope of moving humankind in the direction of a more sustainable way of life. Given the inherent connection between what is considered true or trustworthy and the values at stake in urgent problems, the development of vision is an essential aspect of this knowledge complex. This is the subject of the final chapter of this book.

Questions:

- What are the guiding principles of a complexity-based, integrative research approach?

- What does the co-design and co-production or co-creation of knowledge imply?

- What arguments can you think of in favour of involving non-academic stakeholders in research projects concerning 'wicked' problems? Can you also think of any counterarguments, i.e., reasons why *not* to involve them in research processes?

- How is scientific progress achieved according to Popper?

- And what is the definition of scientific progress for Thomas Kuhn?

- Do you agree with Kuhn that one paradigm cannot be compared to another? Put differently, why is it impossible to measure paradigms independently, i.e. in a theoretically neutral language, with each other according to Kuhn?

- Why is Kuhn seen as a relativist? Is this also true of Popper? If so, why? If not, why not?

- Critics state that scientific progress is no longer possible in Kuhn's vision of science and that it has lost its rationality. Why do they say this? How did Kuhn defend himself against this criticism?

- Why can science no longer be viewed as an internally regulated, rational learning process according to Kuhn? And what is Lakatos' opinion?

- What do you think after reading this chapter: is science making progress, or is it just a matter of continuous development? If the former, how is it possible to determine that science is indeed making progress?

- What does Fay call the new form of objectivity linked to the epistemological concept in which perspectivism combines with fallibilism? Explain why he calls it this.

- Apart from researchers having to examine their research method carefully, they should also be aware of their position in society and be prepared to give an account of their choices. What does Fay mean by this and why is this important?

- Which five regimes of justification does Verschoor identify?

- In what sense do Verschoor's regimes of justification conflict with traditional scientific criteria such as objectivity and value neutrality?

- Besides the methodological implications, what are the societal implications of the implementation of complexity thinking?

- What do you consider to be your task as a future researcher when it comes to solving complex problems?

6 The Future of Science

6.1 Science and Futures Thinking

The subtitle of this book promises that it is not only about philosophy of science but also about vision development. It is now time to live up to this promise. We leave the terrain of 'pure' scientific knowledge and tread into the field of visionary thought, although the idea that science and vision are distinct, separable categories is also scrutinised here. What we hope to show in this chapter is how vision-led science can inform and inspire evidence-based visions that can help safeguard a sustainable future for all who live on this planet.

In the first section, we reflect on the role of science in society today and in the future. After all that has been said, the rationality project – of which science is a part – is critically examined once more. This leads us to conclude that while science has brought us valuable knowledge, it leaves us empty-handed as regards the insights and skills that are needed to put scientific knowledge into practice. We review how the current academic system has helped to discourage the transfer and implementation of knowledge in society, and what may be done to improve it.

We continue by reviewing various ways to implement futures thinking in scientific practice, but not without first investigating what we mean by 'vision'. While we herald inspiring visions that stimulate us to take action and to design potential solutions to the challenges we face, we should avoid the trap of modernity, when 'Vision' was still written with a capital 'V'. In the end, such grandiose 'Visions' often turn out to be disillusions. One way to find inspiring visions may be to develop scenarios that offer us ideas of more or less desirable futures. If a vision is powerful enough, it may even lead to a change in our worldview and trigger a paradigm change. We know that a new outlook on reality that is captured in complexity thinking is thought to lead to (or already have led to) a system change in the field of science. Here we investigate whether this change remains limited to the scientific research practice or whether it is also needed or actually happening in science education.

In the final section, we bring our findings to a conclusion. We claim that, if we want an education system that is fit for the future, we need to go from simple to reflexive modernisation, i.e. to a science and society that regularly reflect critically on the assumptions on which they are based. And we argue that if we want to encourage a truly clever, wise way of thinking and acting, we need to develop a concept of rationality that transcends the level of individualistic interests and embraces planetary thinking. The hope is that if we succeed in broadening our conception of science and society to enable it to accommodate complexity thinking, and if we expand on our conception of rationality to enable it to accommodate not only cognitive knowledge but also valuable wisdom, we may succeed in tackling today's complex problems, thus ensuring a future for humanity on this planet.

6.1.1 Combining Know-What with Know-How

We saw in chapter 4 that the rationalisation process that began two centuries ago actually has two sides. On the one hand, it is the medium in which cognitive-instrumental questions are addressed in an efficient way, and it enables many of us to live a comfortable life. At the same time, the limited 'rational' view of life and the events and challenges it brings with it has serious drawbacks in the form of unintended and unwelcome side effects. The negative effects of human intervention in the natural world is just one – albeit persuasive – argument here. These are the boomerang effects of the Anthropocene age, of our far-reaching changes to the planet's system.

Nicholas Maxwell (2008) argues that science offers improved knowledge – 'know-what' – as well as technical 'know-how'. But it does not provide a form of research enabling people to learn how to deal in a rational way with the growing problems we face, even though, paradoxically, the developed scientific knowledge and technologies are perhaps the main causes of our present global problems. Maxwell considers it a major philosophical blunder for the academic world to have established a form of research that is not aimed at improving the world but is instead focused on the acquisition of scientific knowledge – a blunder that is causing widespread devastation on the planet. To quote Maxwell:

> 'We need a new, more rational kind of academic inquiry that gives intellectual priority to tackling problems of wise living over problems of technical knowledge.' (ibid.: 13)

Edgar Morin (2008: 1-6) is also critical of science's current record. He recognises that the Enlightenment and the advent of rationality helped dispel mythical ideas and provided us with an amazing store of knowledge about physics, biology, psychology, sociology, etc. And he agrees that science continues to expand along empirical and logical lines. Yet the world is still full of ignorance, misapprehension and blindness. He blames ignorance largely on the results of naive ideas about the development of science. The flaws in our thinking do not pertain to facts, in his view; it is not about incorrect observation or logical incoherence (note that Morin was writing before the post-truth age some say we have now entered). Rather, they result from the way we establish knowledge in systems of ideas, theories and ideologies. We still mainly organise knowledge in ways that fail to take account of the complexity of certain issues. Morin also notes that there is a certain refusal to recognise the distorted use of rationality. He fears that these shortcomings will lead to an uncontrollable 'advance' of science with unforeseen and unwelcome consequences such as imbalances in our socio-ecological system.

The acknowledgement of the two sides to the rationalisation process enables us to look beyond the one-sided view that otherwise focuses entirely on the looming dangers. It is true that the rationalisation process has brought far-reaching changes that can also lead to unintended side effects such as a sense of alienation or to dominance, such as the dominance of cognitive-instrumental rationality over other forms of rationality, the dominance of human beings over the rest of the natural world or the dominance of one group of people over another. However, these changes are also the seedbed for liberation and emancipation, for example liberation from the daily grind to supply what we need for our material (re)production, or liberation from socially oppressive situations. We must not forget that the rationalisation process also contains a promise of improvement, emancipation and freedom.

It must be said, however, that, to date, the advantages have fallen primarily to 'the happy few'. Moreover, it seems to be the many who have not benefited who are paying the price of the negative side effects of rationalisation. To correct this biased impact of the rationalisation that has characterised the modernisation process, we must broaden our horizon and examine other aspects that are also essential if we are to live a truly rational life. Science with vision is the way to achieve this broader view. What that means in practice is discussed below. But let us first examine the actual role of science and the potential role that science could play in the optimal exploitation of our rationality potential.

6.1.2 How to Solve the Knowledge Paradox

Chapter 4 explained how the mode 1 approach to science assumes that knowledge produced in universities will find its own way as it trickles down into society. This assumption, referred to as 'pipeline thinking', has not always proved to be entirely accurate: a huge gap still separates universities and society. In Europe, this has produced a so-called 'knowledge paradox': we produce knowledge, but often we fail to transfer it to society (Commissie Toekomstbestendig Hoger Onderwijs 2010: 18; see also Canton et al. 2005: 17).

One possible explanation for this paradox is the way scientific knowledge production is organised, which can hardly be said to be focused on implementing knowledge in society. The production of knowledge is almost exclusively defined as academic 'output'. Applicants to research funds pay lip service to the idea of social valorisation by adding a compulsory paragraph about how the results will be disseminated in society. But the decisive criterion for scientific knowledge production is publication of the findings in highly specialised international journals. The quality and quantity of publications are a major indicator for the distribution of national, international and internal research funds. The focus on output in the form of production of articles in (major) scientific journals and on numbers of graduates has resulted in enormous rivalry in the academic world (Nowotny, 2016, p. 115). The reduction of public money for the universities will only increase the competition. So we see that, just like any other field, science is fuelled by economic and marketing principles.

This has led to a 'publish-or-perish' culture in the academic world. The achievements of scientists are assessed by a short-sighted type of bean counting that involves compiling and evaluating publications. This is coupled with the so-called 'impact factor' of particular journals, a measure of the esteem in which the publication is held in academic circles (ibid.: 157; Dijstelbloem et al. 2013: 9). The impact scores can be decisive in a researcher's career. The research published in these journals may be excellent in theory and method, but its impact and importance to society are a different matter. The highly specialised international journals are rarely read by more than a few dozen or perhaps a few hundred colleagues; for the average reader they are probably indecipherable and definitely far too expensive.

A group of scientists who have launched a campaign called Science in Transition conclude that the scientific system is weighed down by perverse stimuli and systemic faults. They argue that this needs to change.

'Over the next few years, science will have to make a number of important transitions. There is deeply felt uncertainty and discontent on a number of aspects of the scientific system: the tools measuring scientific output, the publish-or-perish culture, the level of academic teaching, the scarcity of career opportunities for young scholars, the impact of science on policy, and the relationship between science, society and industry. The checks and balances of our scientific system are in need of revision.' (www.scienceintransition.nl)

The initiators of Science in Transition argue that science must be appreciated for the value attached to it by society and stakeholders in society, and that these stakeholders should therefore participate in decisions about knowledge production.

A similar ideal lies at the root of the Science Commons project, which builds upon the Creative Commons movement (see Rifkin 2014). It is a movement of scientists who are setting up open-source networks and who are campaigning to make research results freely available. They also protest against the copyrighting and patenting of new insights and technologies since that impedes the sharing of information, stops research, discourages cooperation between scientists and blocks new innovations (ibid.: 180).

The Science Commons initiative may be praiseworthy, but it is doubtful whether it will solve the knowledge paradox. To be able to do that, we need a more profound vision of how best to implement knowledge. A first step is acknowledging that valorisation implies more than just the transfer of knowledge. Scientific knowledge can only become robust social knowledge when knowledge acquisition in society is transparent and inclusive. Helga Nowotny (2016: 6, 83-84) states that the time is ripe for 'citizen science': participation by citizens should no longer be limited to an optional dialogue; citizens should have a genuine say in knowledge production. In some forms of research, such as participatory action research and transdisciplinary research, the co-design and co-creation of knowledge are actively encouraged. Yet these forms of research are more the exception than the rule: in today's conditions, they have low status and less funding.

To realise citizen science, it is important that the public knows how science works, and by the same token it is essential that scientists know how various stakeholder groups and the wider public act and respond. This mutual understanding should preferably be inspired or driven by a vision that can be explained and is comprehensible for a broader public. That vision is process-focused; there is no blueprint for the way this ought to be (Nowotny at al. 2001: 248-249). So what does 'having a vision' actually mean? And how is a vision different from scientific knowledge? When is a claim to knowledge no longer science and when does it become vision? That is an interesting question worth examining further.

6.2 Vision-Based Science and Science-Based Visions

6.2.1 Vision with or without a Capital 'V'?

Let us start by taking 'vision' simply to mean 'a certain perspective on matters'. This is a narrow understanding of the term which states no more than that we always have a particular view of whatever we look at. We know that our perspective influences the direction we take when analysing and studying problems. How we define a problem, how we determine its boundaries, and what type of solutions we devise are all to a large extent influenced by the perspective we take. Imagine, for example, how our perspective impacts at what point we consider a breeding technique to no longer be 'natural' or at what point we believe a crop improvement is 'artificial'. We also know that science is based on certain fundamental assumptions that provide a substantive impetus to the knowledge acquisition process, although these assumptions themselves cannot be scientifically tested. Such assumptions are inevitable because we need a theoretical perspective to turn a research project into a focused and meaningful learning process. To paraphrase Einstein: 'Science without vision is blind'.

At its core, then, science and vision are hardly separable, if at all. The perspective or vision that a scientist takes determines to a large extent the weight attached to the various arguments and the persuasive power with which a particular point is made (cf. Winsberg 2010: 94-119). At the same time, the idea of vision can also be seen in a different light, as implying more than just taking a certain perspective. A vision can also be viewed as a captivating package of future-oriented, scientifically founded, value-charged and preferably also politically well-wrought insights and ideas, assembled in an overarching framework. This vision is best presented with an inspiring narrative or a persuasive metaphor to appeal to a wider audience and to generate broad support for the normative component that shows the direction in which the vision is heading.

The Power of Metaphor

A vision is even more powerful when presented in a striking concept or with an expressive metaphor. For example, GMOs can be labelled 'Frankenstein Foods' or genetically modified corn can be depicted as a pistol-shaped cob. Both are examples of negative metaphors that conjure up strong emotions.

More positive associations are aroused with the term 'Green Revolution' or, more recently, 'Evergreen Revolution' to refer to a farming systems approach that looks beyond the field and even beyond the farm gate. In this systems approach, the farm is considered part of a larger ecosystem (Fresco 2015, 2012: 409). Its aim is to close the cycles of nutrients, water and chemicals, thus creating new models for production and consumption (the 'circular economy'). The slogan 'Waste is food', that is the central focus in the Cradle to Cradle movement – which is itself a gripping metaphor – is used in a similar way (Braungart & McDonough 2002, 2007).

Figure 6.1 Waste is Food

But perhaps this may lead to overly grandiose ideals, which has been precisely the criticism levelled at the monumental concepts of modernity. For two centuries, from 1789 to 1989, from the French Revolution to the fall of the Berlin Wall, Grand Narratives formed the guide and inspiration of many a social movement. But since it has become clear that normative opinions are historically formed and therefore more or less coincidental, and that they may lead to dangerous totalitarian ideologies, these grandiose ideals were dispelled and cast away. Absolute moral truths in the form of ideologies are now forever suspect, because it is impossible to identify a definitive just cause, a greater good that the whole community can support (see Rorty 1989). Consequently, present-day politics is bereft of grand visions. Government does not show any leadership but has become mere management and administration (Harari 2017: 438-439).

This sceptic stance towards grand visions reminds us of the postmodernist attitude that some scientists have adopted. Given that some of the big political visions of the twentieth century led to megalomaniacal regimes with disastrous outcomes, there is something to be said for this scepticism. Playing the devil's advocate, Yuval Harari (ibid.) suggests that maybe we are better off if we leave the important decisions in the hands of petty-minded bureaucrats and the free market. But the danger is that such an attitude may block every initiative to take action. For if we step back definitively from any form of moral truth, it is no longer clear how we can justify ideals such as freedom, equality and solidarity. There is no single inspiring narrative anymore; there are no irrefutable truths making it worthwhile to stand for something, let alone to stand up for something. And if we leave everything to the market, which hand is both blind and invisible, it may fail to do anything at all about the global problems that are threatening us. We do need some kind of grand visions that inspire us to tackle those grand challenges.

The challenge for modern leaders is to project a vision, an inspiring story, without making it a grandiose narrative in the modern sense. In other words, how can they present motivating ideals without falling into the trap of pretending to be the guardians of the absolute moral truth?

Richard Rorty explains that this dilemma can be solved by adjusting the theory to the narrative (note: no capital 'n'!) via the imagination. By this he means that while modern people know (or should know) that their ideas are not universal and eternally valid, they nevertheless try to encourage solidarity among others by envisaging why they want to back or support a cause (ibid.: 189-199; cf. Jasanoff 2017). In short, they present visions – not Visions with a capital 'V' – and try to persuade other people to support their view. They do this by creating a new 'moral sentiment': by making people sensitive to the suffering of others – the suffering of a people somewhere else on the planet or the suffering of animals, for example. Films such as *An Inconvenient Truth* by Al Gore (2006) and Leonardo di Caprio and Fisher Stevens' *Before the Flood* (2016) are powerful examples of this approach. Unlike these perhaps somewhat pessimistic documentaries with their warnings and concerns, others picture more optimistic visions of our future in videos such as *Pandora's Promise* by Marc Lynas (2013) and the *Ecomodernist Manifesto* TED talk by Rachel Pritzker (2016).

Yet another way of introducing inspiring stories is by developing future scenarios. In chapter 2, we promised to come back to the question of how simulations can function as a means to bring virtual realities and possible futures into view. That is what we will do now in the following section.

6.2.2 Virtual Realities and Possible Futures

In addition to supporting theory construction, testing existing explanatory models and exploring new theories, simulations can also serve as a way to 'look into the future' (Gilbert & Troitzsch 1999: 4). As such, they are comparable to – or part of – future scenarios. Scenarios do not offer predictions for the future but rather yield a landscape of possible futures (Bankes et al. 2001: 73). Good simulations can function as computationally assisted thought experiments about how emergence can play out in closed systems. At their best, models allow us to develop insights into and ideas about thresholds, tipping points and feedback or cascade mechanisms and thus help elucidate the emergent dynamics in the complex systems in which our 'wicked' problems are embedded (Törnberg 2017: 61).

Generally, scenarios include a definition of problem boundaries, a characterisation of current conditions and processes driving change, an identification of critical uncertainties and assumptions on how they are resolved, and images of the future (Swart et al. 2004: 139).

Like any other scientific research method, scenario development usually starts with a thorough problem analysis: what are the challenges that we face, maybe not immediately but perhaps in the near or further future? Next, various trends are investigated, particularly those expected to have a huge impact on future society and those that are highly uncertain (i.e. it is uncertain whether they will only set in and then quickly die off again or whether they will indeed hold and intensify). The trends that are deemed most important – i.e. those that are expected to increase exponentially – are chosen as the central drivers of the scenario development. They supply the X and Y axes against which four possible scenarios can be plotted. Both axes represent two extremes on either side of the pole (e.g. scarcity versus an abundance of energy, or increase versus decrease in population). If simulations are run on the four possible scenarios, the parameters would vary according to the position of the scenario on the axes.

For each quadrant, a sketch can be made of the expected state of technology, demography, culture, ecology, and political and economic situation. One of the sketched possible future worlds is the worst-case scenario. But there may also be a preferred future world. This preferred scenario can be taken as the starting point for what we want to achieve. Via backcasting – reasoning backwards (which is the opposite of prediction) – the necessary subsequent actions and interventions can be designed to try to realise this particular future. Road maps can help to build bridges between the present and the future by translating a shared vision into a concrete action plan.

Simulations can be used as a way to develop concepts for the dynamics in complex systems. They can also function as powerful metaphors (Törnberg 2017: 62, 72). Preferably, scenarios do both by using the simulations to present well-founded, complexity-based scenarios in the form of coherent and plausible stories – told in words and numbers and captured in telling titles – about the possible co-evolutionary pathways of combined human and environmental systems.

While our thoughts and actions are usually constrained by existing ideologies and social systems, scenarios can help expand our horizon and enable us to think in far more imaginative ways about our future which in turn may cause us to act differently (Harari 2017: 461). As an example of how to envisage such future scenarios, take a look at the quadrant in figure 6.2. It depicts visions of the future held by acknowledged scientists, thinkers and world leaders.

Futures Thinking: Will We Be Able to Feed the World in 2050?

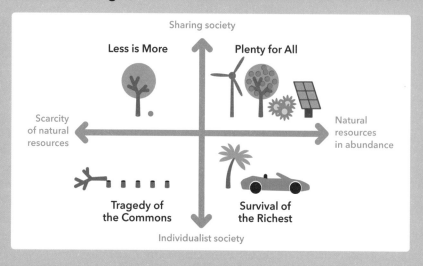

Figure 6.2 Future scenarios: Four possible worlds in 2050

In the worst-case scenario entitled *Tragedy of the Commons*, after Garrett Hardin's well-known article from 1968, there will not be enough resources to feed the future world population. It refers to the situation in which: 'Each man is locked into a system that compels him to increase his herd without limit – in a world that is limited' (ibid.: 1244). In Hardin's parable, the story ends in a catastrophe when each individual herder has put so many sheep on the communal pastoral greens that the grounds become exhausted. The message for modern times is that if we keep increasing consumption without bothering about the limitations of our planet's resources, we may end up with a barren planet (cf. Diamond's *Collapse*, in which he reconstructs how Easter Island became a barren island; Oreskes & Conway's *The Collapse of Western Civilization*; and Wallace-Wells' *The Uninhabitable Earth*). Although these stories are often dismissed as 'alarmist' and such visions of the future are discarded as 'doomwatching', many experts agree that we should take action now to adjust the ecological imbalance. Otherwise, they fear that the system will soon reach a tipping point, a point of no return that might result in a situation that is far less comfortable and perhaps even unliveable for people (Capra 2001; Webster 2007: 40; Barnosky et al. 2012; Gore 2013: 142, 280-282, 374; Rifkin 2014: 10; Byrne 2014: 30). The crises may reach well beyond the ecological system and extend to our cognitive system as it is affected by threats of disintegration, disorganisation and cognitive dissonance. Moreover, there is a danger that our technical systems may break down if they malfunction for too long. This in turn could lead to impasses and deadlocks or chaos and anarchy in our social systems (Rescher 1998: 183).

In the 'Less is More' scenario, there are also serious concerns about our natural resources. Given that the economy is no longer defined as the science of the distribution of scarce resources but rather functions as a system whose main imperative is the maximisation of growth, the fear is that society now runs on a set of suppositions that cannot be maintained in the long run. This will sooner or later result in a systemic crisis, particularly if the world population keeps increasing. The reigning vision in which the socio-ecological system is viewed as part of the socio-economic system and in which the dynamics of our geosphere, atmosphere, hydrosphere and biosphere are merely regarded as externalities of our economic activities has come under intense scrutiny by many acknowledged scientists. They argue that the opposite is true and that we should therefore change our worldview into one in which the socio-economic system is seen as part of the overarching socio-ecological system (Meadows et al. 1992; Daly 1996; Clayton & Radcliffe 1996; Brown 2001). The idea that human well-being can be disconnected from environmental destruction (*decoupling*) is problematised by many thought leaders (see Diamond 2005; Wijffels 2012; Fresco 2012: 380, 476, 481; Gore 2013; Ockels 2014; Rifkin 2014; Visser 2015). Echoing the central message of the Club of Rome's seminal report *The Limits to Growth* (Meadows et al. 1972), they stress that our world is limited, with limited resources – at least as long as no other habitable planets are found. In order to safeguard a comfortable existence, some propose that we transform society and develop a so-called circular economy in which recycling and re-using products is the default, and where mutual services and redistribution replace ownership. Decentralisation is a key word in this scenario. With regard to food production, hope is drawn from the potential of growing organic crops (not GMOs), eating locally produced food and preferably consuming a vegetarian diet.

In the 'Plenty for All' scenario, there is considerable optimism among the visionaries who believe that we stand on the brink of a new era. They think that, under the influence of developments within the socio-economic domain and innovations in biotechnology, nanotechnology, communications technology and computer technology, a third industrial revolution is underway (Braungart & McDonough 2002, 2007; Rifkin 2014: 110, 268-269). Some see the developments pointing in the direction of a circular economy and a sharing, more community-based society (ibid.: 252-253). The 'new economy' is thought to be so fundamentally different from the conventional linear economy, both in its overarching vision and in the underlying assumptions, that it is considered unfeasible to embed it in the current individualist, capitalist system. Some expect the new model to bring the conventional system to a crucial tipping point some time soon. Others actively try to bring an end to capitalism, which they view as an unjust and perverted system (see the *Accelerate Manifesto* of Williams & Srnicek 2013). They want to push towards a future that is more modern and try to unleash a movement towards a more global rational society, towards a completion of the Enlightenment project of self-criticism and self-mastery.

Not all optimists, however, think that a complete system change is needed. A group of scientists who call themselves 'ecomodernists' think that it is possible to turn the Anthropocene era into a good age, a successful period for humankind without huge systemic changes (Asafu-Adjaye et al. 2015; Visscher & Bodelier 2017; see also Gore [2013] who, although not a self-acclaimed ecomodernist, in effect holds the same views). They criticise what they see as a romantic but false and ineffective manner of 'living in harmony with nature' and instead call for a modernist approach. Labelling themselves as grandchildren of the Enlightenment (ibid.: 226-227), they prefer to cling to a rational approach instead of what they regard as an almost religious attitude towards nature. In their Ecomodernist Manifesto, they claim that, if we use our growing social, economic and technological powers well, we can make life better for people while stabilising the climate and protecting the natural world. Decoupling is key to achieving this ideal. In their vision, with agricultural intensification and the efficient use and appropriate management of natural ecosystem services, we need not fear a future where we run the risk of lacking sufficient agricultural land for food or any other natural resources for that matter. They do not believe that the world population will keep increasing, and even if it does, they think human beings are inventive enough to find solutions to scarcity problems. By committing to the processes that are already underway, they trust that an ecologically vibrant planet can be achieved in the near future.

But critics (e.g. Piketty 2014; Klein 2017; Hajer 2017) fear that this abundance will only be available for 'the happy few', and here we come to the fourth scenario, which some alternatively label 'Survival of the Richest' (Osnos 2017). With regard to the food issue, it may be noted that hunger and malnutrition posed a huge challenge in developing countries in the twentieth century. And even now that we have entered the twenty-first century and obesity has become the biggest problem globally, problems of hunger and malnutrition remain unsolved. Considering the continuous innovations in food science in recent decades, it may still be possible to complete the 'Green Revolution' and increase food availability worldwide

At the same time, it is fair to raise the question whether science can indeed fulfil this promise. From the start, it was clear that there were environmental costs such as morbidity and mortality resulting from an excessive use of pesticides and water pollution, as well as social costs in the form of farmers who were banished from their land. And still, new innovations, food. technologies as well as many other innovations seem to predominantly benefit people in countries that are already well-off, while those who are in need remain vulnerable. Some visionaries warn us that it might very well be that in the not so far future, these vulnerable groups may come to be regarded as 'redundant' by Artificial Intelligence and the privileged elite of genetically engineered and technologically upgraded 'superhumans' that will then dominate the world, which/who may decide to 'let go' of what they regard as a useless, inferior caste (Harari 2017: 356-411).

The notion of alternative futures and the recognition of the possibility of progress informed by reason is an important – and to some even crucial – asset in the implementation of complexity thinking. Byrne, for instance, is committed to the latter belief. He argues that the recognition of alternatives and the role of purposeful action in achieving different alternatives is precisely what distinguishes complexity thinking from more traditional, structural accounts of how reality will develop (Byrne 2014: 141-145). Scenario development can help illuminate the alternatives as well as the different ways in which things might be done to achieve the desired sorts of futures. As such, it is a useful method in all scientific fields concerned with 'wicked' problems, including Byrne's own field of planning, which he views as 'almost the archetype of modernity as process' (ibid.: 141). Extrapolating his view on the importance of complexity thinking to the broader scientific field, we could say that complexity-based scenarios provide a rational framework for thinking about and mapping out possible futures – a framework that is not based on simplistic determinism but rather is explicitly founded on reflexive social action. Although it breaks with modernity's outdated idea of a manageable society, it does offer us the possibility of making rational choices by showing the condition space that defines the possible scenarios, thus enabling us to take a path in the direction of the future we want.

6.2.3 Visions as Forecasting Paradigm Changes

It is clear that the search for explanations and solutions to urgent problems is directed by a particular vision. It does not matter whether we view visions in the narrow sense of 'departing from a particular perspective' or in a broader sense in which visions serve as guiding scenarios for a future world. In both instances, visions are related to particular paradigms. We have come to know paradigms as the overarching frameworks that guide research programmes of groups of scientists. But the use of the concept 'paradigm' has not remained limited to the domain of science. Over the course of time, the term has been extended to include not just theoretical frameworks within science but any overarching thinking system or worldview. Within the systems thinking framework developed by Meadows et al. (2008), paradigms and paradigm changes represent the most influential *leverage points* for interventions within a system (see figure 6.3).

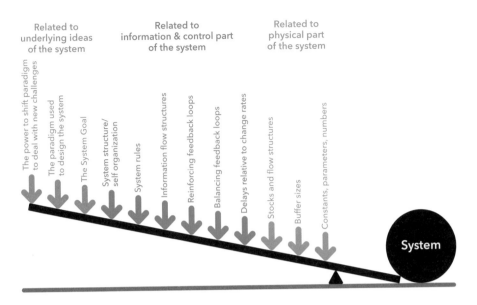

Figure 6.3 Paradigm shifts as most influential leverage point for system intervention
Source: Meadows 1999; Meadows et al. 2008

A shift from one paradigm to another occurs when a whole new perspective, a fundamentally different vision of reality is adopted. The previous chapters have given us examples of some fundamental shifts or turns that have taken place within science. The question of whether the introduction of complexity thinking has led to a paradigm change is something we have dealt with extensively in this book. This chapter has shown that paradigms can also be very influential with respect to the future vision of society as a whole. Many acknowledged scientists and leaders hope that a change in the dominant vision will take place and that this will lead to a paradigm shift. They feel that intervention in the highest reaches of the system is urgently required because dealing with the symptoms at the lower levels is not going to help solve our problems.

One way to approach these systemic changes in a positive way is provided by transition thinking and transition management. Transition refers to structural changes brought about by new ideas, technologies, products, frameworks, discourses, infrastructures, practices and patterns that result in the emergence of new systems (Grin et al. 2010: 18-28; Spaargaren et al. 2012: 4-6). The idea is that innovations emerge bottom up, from the small *niches* of a system (see figure 6.4). They often follow new inventions and new technology. Where existing day-to-day practices can adapt sufficiently and permanently to new directions (*deroutinisation*), the niche developments eventually result in a new socio-technological *regime*. This is systemic change. When enough stable new patterns in turn form within the regimes and these are embedded and institutionalised, it may result in a paradigm change. A transition may be encouraged or hindered by the overall *landscape*, i.e. by major external societal conditions. A development in the opposite direction is also possible:

when major adjustments occur at the landscape level, a top-down change can be implemented. For example, if worldwide energy prices were suddenly to increase, this would create new opportunities for niche innovations for sustainable energy to develop into new regimes (ibid.: 7-12).

An advantage of transition thinking and transition management is that it views the agency of social actors and societal structures as mutually dependent and in constant interaction with each other. This approach deals with the duality of structure and emphasises that the potential for change (or obstruction of change) can be found both on the structural side (i.e. factors pertaining to the system) and on the action side (i.e. factors pertaining to the actors in the system) (Bos & Grin 2008: 483; Grin 2012: 36-37). As a result, it is possible to avoid the kind of unfruitful dualisms that hamper an adequate, integrative research approach to complex issues. Structural changes are necessary to achieve a breakthrough when a system is in deadlock. A system may be significantly constrained by lock-in mechanisms that block or conceal alternative visions of significant aspects such as infrastructure and users' routines (Grin et al. 2010: 19-21; Spaargaren et al. 2012: 4-6, Van Mierlo et al. 2010: 35). But a good vision has the power to give new regime elements a chance, not least because it offers the inspiration and sustenance needed to generate a constructive interaction between the regime elements and the innovative practices that launch the change (Bos & Grin 2008: 499; Grin 2012: 54).

Below, we examine the implications of the proposed paradigm shift in science and society for the vision of academic education, which is supposed to prepare students for a career as a researcher or an investigative professional in the twenty-first-century knowledge society. After addressing the question of whether such a paradigm shift is indeed underway, we also examine various action perspectives that may help induce change in the educational domain.

6.2.4 Transformative Learning as New Educational Vision?

The discipline-transcending paradigm offered by complexity thinking can do justice to the many interactions that take place at the interface of humanity and the planet. That is in itself necessary because attention for the importance of those interactions is often absent. This is the argument of the researchers at the International Council for Science (ICSU) who dedicate their efforts towards international scientific cooperation to meet the 'grand challenges' in the world today:

> 'One implicit assumption relevant to education and capacity building in response to global environmental change may be the dualistic understanding of human-environment relationships that represents the ontological basis for modernity and positivist science (Castree 2005). Although theoretical perspectives seeking to transcend this dualism can be traced to both exact sciences (Von Bertalanffy 1969; Prigogine 1977) and social sciences (Murdoch 1997; Manuel-Navarrete & Buzinde 2010), many approaches to the problems discussed above nonetheless fall prey to 'maintaining a dualistic separation of

ourselves and nature, economics and ecology, subject and object, present and future' (Sterling 2009: 108), and point toward the need for deeper inquiries into the assumptions and beliefs underlying such approaches (O'Brien et al. 2013: 53-54).'

In chapter 5, we presented an approach in which existing dualisms are transcended to launch a knowledge acquisition process that is able to generate robust solutions to our complex problems. But how does this translate to the educational setting? The criticism of today's dominant paradigm in science also has implications for the vision of education. What does this new view imply for the knowledge, attitudes and skills needed to translate new thinking into different ways of acting?

As a subsystem of a larger system, education is designed according to the same logic as the reigning paradigm. It is therefore vulnerable to the same risks of system failure as its overarching system (Sterling 2011: 53-54). Bela Banathy (1993, 1999, 2001), Ken Robinson (2010) and O'Brien et al. (2013) all agree that a new vision and a paradigm shift is needed for education. They see the nineteenth-century education system with its conveyor-belt institutions as fundamentally failing in the post-industrial knowledge society of the twenty-first century. And this problem analysis is shared in the corporate world (see World Economic Forum 2016 which, incidentally, states that we are in the midst of a fourth rather than a third industrial revolution).

Banathy (1993, 1999), Paul Maiteny & Ray Ison (2000) and Stephen Sterling (2004, 2007) claim that a suitable approach requires a fundamental change in knowledge theory, especially for education where sustainability is key. They suggest a shift from a machine metaphor (the university as a conveyor-belt institution) to a vision of learning in which the educational institute becomes a vibrant system with a learning organisation. This suggestion calls to mind the proposal, made in a previous chapter, to compare the way we actually learn and gain knowledge to the functioning of a coral reef. The proposal is inspired by the perceived need to move beyond critical rationalist and interpretivist paradigms and to present the knowledge acquisition process in more relational terms. In this vision, knowledge is not conceived as a commodity, as a thing or as data stored in a researcher's brain or in computers but is understood as a dynamic participatory connection between researchers and the world. Knowledge is not obtained by passive observation but by experimenting and actively intervening in the surrounding environment (see also Midgley 2000: 44).' To say that a person knows something means that their knowledge enables them to interact effectively with their environment.

Viewed from this angle, knowledge is not about 'the world' as such; it is about what we can do in the world and how we can change the world. This fits very well with the plea to gain more 'know-how' and leads Sterling (2004: 54-57; 2011: 22-26, without a doubt inspired by Schön 1987) to argue for transformative learning, which implies learning at three levels. At the first level, that of basic or single loop learning, knowledge is transferred regarding external, physical and social reality. This first level is all about cognitive knowledge and instrumental rationality. The second level is about learning about learning, about double loop learning, meta-learning and meta-cognition. The third level of learning goes even deeper. Here learners reflect on the assumptions they hold, including the used knowledge theory; hence this type of learning is also labelled epistemic learning. Ways of observing and interacting with the world – e.g. frames of reference and the research methods that are used – are critically examined, as are the underlying regimes of justification that direct these actions. Only at the highest level of learning does Sterling (2007) identify the possibility to change the paradigm, though such a paradigm change should be thoroughly prepared and executed at all different levels involved in education (see table 6.1).

Three part world view	Ontology	Epistemology	Methods
Academic terms	Conceptions Cognitive dimension	Perceptions Affective dimension	Practice Inverventional dimension
Everyday terms	Knowing & Understanding 'Head'	Seeing & Awareness 'Heart'	Doing & Competence 'Hands'
Educational terms	Policy & Curriculum Learner is connected	Paradigm & Purpose Learner is concerned	Pedagogy & Practice Learner is capable
Strategic terms	Critique	Vision	Design

Table 6.1   Tripartite model for paradigm change in education
Source: Sterling, 2007

In order to properly execute the suggested transitions in science and scientific education, a consequent shift in the learning paradigm would seem appropriate (Banathy 1999, 2001; Ison 1999, 2001; Grobstein 2007: 26; McCarthy et al. 2011; Wiek et al. 2015). Ken Webster (2007: 42) and Jeremy Rifkin (2014: 113) are positive regarding the paradigm change towards complexity thinking suggested by Morin. They see the reductionist approach to learning (based on the isolation and separation of components, which was very useful in the industrial age) gradually superseded by a more systematically focused learning aimed at understanding subtle connections that link phenomena in large contexts. This would be a good way to connect with the transformation currently underway in the global economy, which the World Economic Forum (2016: 3, 25 and 32) predicts will lead to disruptions in the employment and skills landscape.

Indeed, movements in the desired direction are discernible. New technologies are creating opportunities to adapt education to the demands of a new age, which calls for twenty-first-century learning and skills. They offer tools to acquire higher cognitive competence, deeper understanding and higher-level learning and thinking (Anderson 2011; Wang et al. 2013; Wallace et al. 2014; Braseby 2015: 259). Here and there, attempts are being made in classes and lecture rooms to abandon the authoritarian, top-down educational model in favour of a more collective learning process. The teacher's role shifts from schoolmaster to facilitator and coach, for whom updating critical learning skills and attitudes is at least as crucial as conveying state-of-the-art knowledge.

Yet it seems doubtful that a real transition is taking place in education or that a 'genuine revolution in pedagogy' will happen, as Rifkin (2014: 109-110) assumes. The fundamental systemic changes needed to develop complexity thinking and interdisciplinarity have yet to materialise. In the present constellation, universities are still mainly focused on delivering monodisciplinary trained scientists. During their studies, academics gain little training or experience in managing complexity and implementing scientific knowledge in society (World Economic Forum 2016: 25, 32), nor are they made familiar with the obstacles they may encounter or the unintended and unwelcome effects that may accompany complex issues (cf. Van Merriënboer et al. 2002: 40, 2009; Wieck et al. 2011, 2013).

For pioneers in interdisciplinary research and education relating to complex issues, it is difficult if not impossible to fall back on an 'outside-in' approach, for there are still only a few interdisciplinary researchers available to consult. Instead, pioneers may have to take an 'inside-out' approach (see Barth & Michelson 2013; Barth 2015: 24-25): they may try and devise ways to encourage interdisciplinary scientific research on complex issues via innovative education. In other words, we can approach the matter from the other direction: what kind of knowledge and skills do today's students need to be successful interdisciplinary researchers of complex issues? They are tomorrow's researchers twice over: they will examine future challenges, and in ten to fifteen years they will hopefully also be part of the academic corps. Only then will it be possible to combine an 'inside-out' approach with an 'outside-in' approach.

Let us hope that, in the meantime, work will continue on a system based on a clear vision of what contemporary education should provide to students. In order to create a system that does not simply help produce (or reproduce) current education practices but challenges accepted assumptions, a critical systems perspective is needed (see Gregory 1993: 206). In the final section, we investigate how such a critical perspective can be incorporated into a more comprehensive conception of science.

6.3 From Funnel Vision to Comprehensive Science

Innovation requires the adaptation and translation of learning processes designed to generate potential solutions in a complex and dynamic environment. Each system innovation, whether in education or the food sector, involves a combination of technical, social and institutional change. The new challenges require new institutional structures designed to bring our understanding and management of our physical world in line with that of our social world, and to deal with the unintended consequences of human intervention in the world (Nowotny 2016: 84). *Reflexivity* – or the re-examination and readjustment of assumptions in light of new information or knowledge – plays a key role in this process (Giddens 1991: 20). As Paul Raskin (2008: 469) remarks:

> 'The shape of the global future rests with the reflexivity of human consciousness – the capacity to think critically about why we think what we do – and then to think and act differently.'

As it is such a crucial skill for vision development, we will devote the last part of the book to a closer examination of the art of reflexivity.

6.3.1 From Simple towards Reflexive Modernisation

Reflexive modernisation refers to the process in which the basic underlying ideas of the social development process are themselves the subject of discussion and examined for inherent contradictions. This involves a modernisation that takes into account the undesired and unwelcome effects of the rationality project, with the intention of creating a clear perspective on their relevance for present-day society. Ideally, it also leads to ideas and suggestions for possible solutions to those unintended side effects (Beck 1986: 253-254).

At best, reflexivity represents the ability to become aware, in a systemic way, of how the world is changing. This requires an understanding of the complexity of the challenges facing us and of the interfaces that emerge as a result of the co-evolutionary process of science and society. How has society managed to gain knowledge about nature and to learn how to manipulate and control it? And how did society develop as a result of this attitude towards nature? Helga Nowotny (2005: 29) thinks the interface of nature and society is the perfect place to study how scientific understanding, intervention and social relevance are generated by our social actions and by the social structure of science and technology. She argues that these form the two main challenges of complexity: both the continuous increase of complexity and our attempts to reduce it to manageable proportions.

In a critical-reflexive science, existing convictions are not just blindly accepted; they are founded, assessed and legitimised based on criteria connected to a broad rationality concept. This concept cannot just be imposed as a norm onto society, however. We would not want to revert to a paternalist attitude, nor would we want to fall into the dogmatic trap of blindly reproducing existing opinions and traditions. So first of all, a critical-reflexive science calls for a critical stance towards traditions.

To enhance the development of what he calls a reflexive balance method, Habermas (1996: 121-122) invites scientists to approach traditions as expressions of continuing learning processes. A second condition is that, prior to any critical evaluation, critical-reflexive scientists must declare their own position. This is necessary in order to be able recognise the learning processes that have taken place and to assess whether they can be regarded as useful or not. In general, science will have to develop further into an institution characterised by unlimited self-criticism, connecting to the potential embodied in social movements of the same ilk. This way, research can be employed as a supportive instrument in the social debates about important societal choices and decisions. This requires a constant attentiveness to ensure a fruitful integration of common sense knowledge and scientific knowledge, and to ensure that scientific knowledge is fed back into society and used in everyday practice.

John Grin, Francisca Félix and Bram Bos (2004, 2008) introduce the concept of *reflexive design* as a way of implementing reflexive modernisation in practice. This is more than just a meta-narrative and implies more than a smooth process of social change. In their view, it is a way of doing research and attempting to achieve progress inspired by a vision and learning from the past (Bos & Grin 2008: 482). It is a learning process in which assumptions, knowledge claims, distinctions, roles and identities once taken for granted are examined critically in the hope of ultimately arriving at an integrated assessment that supersedes existing restrictive differentiations and distinctions (Grin et al. 2004: 126, 128).

Reflexive Design in the Food Sector

With their book *Food Practices in Transition: Changing Food Consumption, Retail and Production in the Age of Reflexive Modernity*, Gert Spaargaren, Peter Oosterveer and Anne Loeber (2012) aim to make a contribution to the systemic reflection on transitions in the food sector. Transition theory and transition management – both specifically designed to deal with complex and persistent problems – can serve here as theoretical frameworks to study the dynamics of change and to guide change processes in the right direction wherever possible.

After the Second World War, the production, processing and consumption of food in Western countries focused initially on increasing efficiency and continuous (technical) rationalisation of the process. Due in part to undesired system effects that ensued when the food system adopted the form of an industry, this approach was adjusted in the 1980s. This resulted in the emergence of a complex hybrid compromise regime in which economic growth was supposed to go hand in hand with sustainable food production and consumption. Inasmuch as this model was able to function with apparent success for a while, this was because it was possible to compartmentalise external aspects and 'export' these to developing countries, so the analysts argue. But since we only have one habitable planet at present, shifting risks and unwanted side effects to another part of the world cannot be regarded as a durable solution (Marsden 2012: 291-292).

Following the global crisis caused by rising food prices and the volatility of the global food markets in 2007 and 2008, the dynamic of the food industry changed again (ibid.: 291). Yet the food industry still seems unable to cope with the resulting environmental and health hazards or to meet the growing concerns of consumers regarding animal welfare and systemic sustainability (Spaargaren et al. 2012: 1-6). Moreover, the West has attained a level at which food scarcity has been replaced by an overabundance of food. The system is beginning to suffer from 'underconsumption' (Marsden 2012: 292).

Waning trust among consumers in today's food technologies and the expert systems surrounding them has increased pressure on the existing food regime and landscape (Spaargaren et al. 2012: 22). This has had an impact on leading players in the dominant regime, such as labelling and regulation bodies. As a result of this shift, the focus has moved away from food producers towards consumers, who are now the key figures in the debate and in the analyses of food patterns. The slogan 'farm to fork' has been replaced by 'fork to farm' (ibid.: 19; see also Fresco 2009: 384). Yet the power of retail traders should not be underestimated. Following the shift that began in the 1980s from state-supervised policy to retail market corporatism, major retail companies now play a key role in translating consumer demand to producers (Spaargaren et al. 2012: 330-332). The role of food chains and the mutually influential developments at the global and local level (known as *glocalisation*) are key issues in the debate about sustainable food consumption and production (ibid.: 11, 15).

The response from consumers can stimulate niche innovation locally, for example by consistently using alternative providers who only supply local or organic food or food that has not harmed animals. By changing consumption patterns and choosing alternatives, consumers can put their preferences on the political agenda. If enough people are mobilised, they may be able to move the food system beyond a tipping point (Kjærnes & Torjussen 2012: 99; Spaargaren et al. 2012: 329). In that sense, the shocks of 2007 and 2008 may well have opened some 'windows of opportunity' for a profound transition.

Some scientists are optimistic about these changes and predict the emergence of new 'foodscapes' (Kloppenburg et al. 1996; Spaargaren et al. 2012: 6, 333). This may lead to the gradual bankruptcy of the present system. It may ultimately instigate a transition to a more sustainable agri-food paradigm that involves producers as well as consumers (Marsden 2012: 291). But there is no guarantee that niche developments will lead to institutionalisation at the regime level and changes to the food system. After all, there are also developments that seem to be working against such a transition. Think, for example, of the commercialised organic food chains that have lost their connection to the local context and other companies that are taking advantage of 'green' labelling without actually changing their existing practices (Kjærnes & Torjussen 2012: 99-100; Marsden 2012: 301). After various epidemics such as foot and mouth disease, mad cow disease, bluetongue and bird flu, governments responded with a new wave of institutional regulations reflecting a continued reliance on technocratic and rational scientific assessment procedures and practices. According to the analysts, the inherent systemic problems are thereby shifted to retailers; they have become responsible for a privatised regulation of food quality (ibid.: 296-299). Meanwhile, it is the food consumer who bears the risks (ibid.: 327).

Spaargaren et al. argue that we are in the middle of a complex transition process with many facets in which governments play a critical role, since they make the rules for food production and consumption. The big question is whether a combination of government and corporations can guarantee food safety while keeping external ecological and social factors under control (ibid.: 292). Analysts fear that the technically refined public-private model is more likely to lead to periodic crises than to deal with them structurally. They argue that problems and risks relating to food are kept out of sight. At its core, the system is powered not by consumer interests but by an economic model in which the priority is always to reduce costs.

The analysts identify other, more pressing reasons to fear that the crises of 2007 and 2008 were more than temporary problems and that the system has reached a fundamental boundary. It has become clear since then that the system suffers both from risks from within the system itself (endogenous risks) as well as from external risks (exogenous risks), which are inextricably linked to the Food-Water-Energy nexus. It is probably no coincidence that food prices rose at the same time as oil prices rose. The food system is heavily dependent on oil. Two-thirds of the energy used is for artificial fertiliser; the rest is for processing and transport. The production of bio-fuels from crops also played a role; this accounted for an estimated 75 percent of the temporary price rise. In addition, there is the increasing water shortage. More than 70 percent of potable water is used in agriculture. Often water is transferred from areas where water is scarce to areas where water is plentiful. And the nexus can be stretched even further, because the pressure on water is increasing due to population growth and climate change, the latter of which in turn can largely be attributed to the way agriculture and cattle breeding is organised (ibid.: 301-302). All this fuels their fear that the system has become untenable and that the limits of the planet's ability to support the system are in sight.

This analysis raises the question whether the foundations of today's food system are genuinely being reassessed and if modernisation is indeed truly reflexive, or whether we are still only halfway there. Either way, Spaargaren et al. conclude that there are many new food landscapes. Inasmuch as a transition will take place, they consider it unlikely that this will lead to a new total system. It is more probable that global multidimensional – and in some aspects heterogeneous – sustainable food regimes will emerge (ibid.: 331-332). And they argue that this is a good thing; for where several regimes exist, the risk of a cascading effect in a crisis is reduced.

This example shows the role that individually organised reflexive self-projects and individual actions can play in enhancing a movement away from a social order based on a one-sided rationality and unrestricted economic growth (Giddens 1991: 223-224). The existential political issues we face force us to question our half-modern systems. This may lead to a reconsideration of the foundations and the underlying principles on which these systems rest. Moreover, the aspects that individuals consider important, judging from their self-projects, will partly determine the political agenda. By raising such existential political questions, those moral and existential issues that modern core institutions have suppressed or seemed to have 'solved' (cf. the truth funnel), are put back on the agenda (cf. Beck 1986). As Ilya Prigogine (2000: 37) remarks:

> 'The future is not given. Especially in this time of globalisation and the network revolution, behaviour at the individual level will be the key factor in the evolution of the entire human species. Just as one particle can alter macroscopic organisation in nature, so the role of individuals is more important now than ever in society.'

Post-half-modern science is characterised by constant reflection on and critical review of its principles and consequences. To become institutionally reflexive, critical self-examination must be part of the core of every scientific institution. Only when science has turned itself into a learning organisation (cf. Senge 2005) and scientific rationality has become a social rationality can reflexive science be said to exist. As Ulrich Beck (1986: 297-299) comments, this requires a different knowledge theory and rationality theory as well as a different relation between theory and practice. What is needed is a 'learning theory' with regard to scientific rationality that does not treat the risks and undesired effects that science itself produces as immutable. What is proposed here is for social actors to use the rules and resources of the system in a new manner. Since their altered actions will impact the system through the feedback loop that exists between actions and systemic structures (think of the duality of structure), the system will eventually change.

To enhance this process, Giddens (1990: 177-178) proposes a utopian realism in which reflection on held assumptions and openness to change is connected to equally crucial concrete actions. This requires us to work on realising our ideals now, without losing sight of what is needed in the future. This way, we can keep an eye on the 'windows of opportunities' that open up for new developments. At the same time, we must watch out for institutional obstacles or rebound effects that may form a threat to well-intentioned changes. Reflexive forms of research, such as research based on systems thinking (see Meadows et al. 1999, 2008; Kennedy et al. 2018), action research (see Almekinders et al. 2009; Van Mierlo et al. 2010), transdisciplinary research (see Pohl & Hirsch Hadorn 2007; Hirsch Hadorn et al. 2008) and design thinking (see Bruchatz et al. 2018) can fulfil a guiding function here. Set up as a reflexive learning process, these types of scientific learning processes can help monitor complex projects around system innovations. Where necessary, they can also help make adjustments to enable participants to contribute as best they can to the realisation of structural change (cf. Grin et al. 2004; Byrne 2005).

6.3.2 Towards a Super Rationality, or How to Live a Wise Life

There is no need to be a pessimist or defeatist to justify reflecting on the so-called 'tragedy of the commons'. If we want to avert this worst-case scenario, it is useful to have insights into the mechanisms that threaten to drag us into this trap. Garrett Hardin (1968: 1245) points out that as long as we behave as independent, 'rational', free entrepreneurs, we will remain imprisoned in a system of our own making. Tempted by our tendency towards psychological denial, we bury our head in the sand as long as possible and continue to use up nature's resources for our own pleasure, even though we know that society as a whole will pay the price. As we ourselves are part of that greater whole, in the end we will also inevitably have to pay the price.

There are theorists who have good hope that this attitude might change in the near future. Rifkin (2014: 65) seems to nurture that hope at least. He argues that far-reaching economic changes are occurring that will affect our awareness profoundly. He envisions a new economic paradigm accompanied by a fundamental reappraisal of human nature, which will subsequently lead to a fundamental reassessment of the way we relate to the planet. Such a reassessment is definitely needed if we want to escape the threat of the 'tragedy of the commons'.

Philosopher Douglas Hofstadter (1985: 737-780) notes that we can solve the problem – also known as the prisoner's dilemma – by approaching it not rationally but super-rationally. This means that we choose the best option based on the idea that the other will make the same choice. An example of such super-rational behaviour would be that, instead of leaving it to others to bear the responsibility and the accompanying costs to effectively reduce carbon dioxide emissions, we all choose to take action, knowing that we can only tackle the climate problem if we all work together. This is just one of many ecological problems in which the prisoner's dilemma plays a role and which can only be solved by cooperating (Buclet & Lazarevic 2015: 92-93). For this to work, trust and faith in other people's goodness are essential. That sounds easier than it is, particularly when global problems are at hand. It requires a radical change in our thinking and a shift away from our tendency to protect our own turf in favour of a new sense of global engagement and responsibility.

The main moral change here is to aim to live no longer like a homo economicus but as persons who care about the common good. Only then can we hope that the required changes will take place, and that sustainable solutions to complex questions such the climate problem and the food issue will be identified and implemented. This requires a new approach in which rationality is not considered solely a cognitive-instrumental way of thinking but one in which global values are taken into account as well. Such a way of thinking is far from impossible, because humans are not solely focused on self-interest, autonomy or on acquiring more property. Human beings are social animals capable of empathy and compassion, drawing satisfaction from sharing and being good to others (ibid.). Sheila Jasanoff (2014: 95-97) detects hopeful signs pointing in the direction of a more global awareness, in the form of emerging concepts such as the precautionary principle, sustainability and intergenerational equity. And Rifkin (2014: 296-303) notes the development of a 'biosphere consciousness', of an expansion of our empathy to encompass all of humanity, even all other life forms on the planet.

Rifkin seems to appeal to the same qualities that Rorty is counting on with regard to solving the moral dilemma of modernity. This dilemma involves conflicting moral obligations. Its cause is the discrepancy between general moral obligations and our own private interests or those of the relatively small social circle with which we identify. There is no philosophical tribunal (or any other kind of tribunal) that can solve this dilemma and tell us which set of moral obligations is right. We can do nothing but rely on own moral frameworks. Rorty (1989: 193, 197) explains that the dilemma can only be 'rationally' resolved by appealing to the moral idea that higher-order obligations – such as obligations to ensure the continuation of humanity – must be given priority above rival lower-order obligations that only serve private, short-term interests. This will only work when there is a willingness to adjust or expand our own frameworks wherever necessary.

This suggestion brings to mind Verschoor's analysis of the regimes of justification as frameworks with which stakeholders appeal to the general validity of their arguments. And it also connects with the plea to no longer view value neutrality as a kind of objectivity but to link it to demands such as accountability and transparency. Researchers are expected to be open about their frames of reference and the regimes of justification with which they claim validity. That makes it easier to check and criticise their ideas. Making these referential frameworks more inclusive can produce a broadly supported framework of legitimacy. Verschoor's framework, which is based in part on his own immediate experience relating to the GMO controversy in Mexico, gives us an insight into the options available to bring together the different regimes of justification of different groups of stakeholders at the meta-level. This could be viewed as an innovative approach, a functional instrument based on a broad rationality ideal capable of producing a new form of objectivity and neutrality in multidisciplinary, interdisciplinary and transdisciplinary research.

To produce problem analyses and solutions that can help us deal with the complex challenges we face, we need up-to-date knowledge ('know-what') as well as technical 'know-how'. We also need insight into how these technologies relate to people's values and how they can be successfully implemented. Broadly trained interdisciplinary researchers can be of tremendous value in this new context, as they possess both technical knowledge and the practical and social skills needed to bring all the elements together. They are able to contribute to an open innovative exchange of ideas, concepts, methods and empirical findings. Using both cognitive-instrumental rationality (i.e. scientific knowledge and technical knowledge and skills) and social or normative rationality (i.e. knowledge, attitude and skills relating to communication and public opinion formation, and a sense of how judgement and decision-making processes emerge), and affective rationality (i.e. the ability to reflect on and be sensitive to the attitudes, positions and sensitivities among a wider public), they can help realise the transition from the old to a new paradigm (Tromp 2013: 14, see figure 6.4). As Valerie Brown, John Harris and Jacqueline Russell (2010: 47) remark:

> 'In the final analysis, knowledge about the physical (instrumental rationality) and social worlds (practical rationality) can be legitimised only through inter-subjective critical reflection that enables the inquirer(s) to become aware of the ways in which their own purposes, agendas, values and those of the historically situated cultural contexts they find themselves in have influenced the inquiry processes, outcomes and consequences (critical rationality).'

A broad concept of rationality is not enough; we also need a theory of knowledge that offers sufficient space for a broad concept. The traditional science and education paradigm – in which knowledge is a product of an objective, individualist, ahistorical, linear procedure – cannot contain a rationality concept like this. But a paradigm in which knowledge creation is considered an interactive, co-designed process allowing researchers to let their knowledge constructions be critically tested by fellow researchers (i.e. their scientific colleagues and co-researchers in the situation in which the project is executed) can. Within this conception of science facts, normative questions and power relations can all be discussed in an attempt to find the most rational course of action. They can all be topics of research in the process of rational knowledge acquisition that emerges in this mutual learning process.

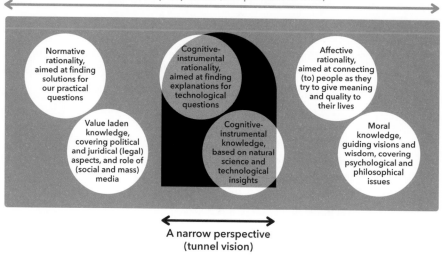

Figure 6.4   A comprehensive perspective on the implementation of scientific knowledge
Source: Tromp, 2013: 13

Such a vision of science and the societal value of science alters the traditional image of the university's role significantly. Instead of depicting scientific knowledge acquisition as a disinterested activity designed to produce generally valid, neutral and objective knowledge, now the university is presented as an institution in which scientists take account of the way they acquire knowledge and of the values that are inevitably connected to the knowledge process. Although extremely complicated, this is the task that scientists who deal with complex problems face. For it is only when we acknowledge the actual complexity of these issues that we can avoid becoming trapped in tunnel vision. An expanded vision of rationality requires us to abandon the limited traditional framework and to adopt a far wider, fuller perspective of science and society. Only then will we be able to develop a truly rational science that can generate adequate future-oriented visions and solutions and can help resolve the complex challenges at the interface of humanity and the planet.

Questions:

- What is meant by the statement that the rationalisation process has two sides?

- What does the 'knowledge paradox' entail, and what is its cause?

- What is the difference between a Vision with a capital 'V' and a vision?

- In what way can scenarios perform a supportive function with regard to vision development?

- To what extent do you consider a broad rationality concept essential for vision development?

- Can you relate vision development to quality criteria? If so, what criteria should these be?

- What connection do you see between vision development and systems thinking?

- What is the role of niches, regimes and landscapes in the Multi Level Perspective on transitions in society?

- Sterling thinks that a paradigm change occurs only at a certain level of learning. What level is this, and why does he think this is the only level at which paradigm changes can occur?

What do you think after reading this book:

- Is a paradigm shift required in science, or is such a shift not necessary?

- And what do you think of education; is a paradigm shift needed in this domain?

- And in society as a whole, do we need a transition?

- What is your view on science now at the end of this book?

- What role does the concept of rationality play in your view?

- And what does vision development mean to you?

- How do you see the role of scientific knowledge in vision development?

- In what way can complexity thinking contribute to a more reflexive modernisation, i.e., to a modernisation that takes account of the negative effects of the rationality project that began in the Enlightenment and that tries to put forward solutions for those unwelcome side effects? (If you wish, you can take the global food crisis as a concrete example.)

- Are you optimistic or pessimistic about the future? Put differently: if you were to take a position on one of the future scenarios sketched here, which would you choose?

- Regardless of which scenario you have chosen, do you see any possible strategies or ways to act in order to achieve a sustainable future?

References

Adebesin, F., Widhalm, J.R., Boachon, B., Lefèvre, F., Pierman, B., Lynch, J.H., Alam, I., Junqueira, B., Benke, R., Ray, S., Porter, J.A., Yanagisawa, M., Wetzstein, H.Y., Morgan, J. A., Boutry, M., Schuurink, R. C. & Dudareva, N., (2017) Emission of volatile organic compounds from petunia is facilitated by an ABC transporter, *Science*, vol. 356, no. 6345: 1386. DOI: 10.1126/science.aan0826

Almekinders, C., Beukema, L., Tromp, C. (eds) (2009) *Research in Action. Theories and practices for innovation and social change*, Wageningen: Wageningen Academic Press, pp. 185-205.

Anderson, L.W. & D.R. Krathwohl et al. (2001) *A Taxonomy for Learning, Teaching, and Assessing: A Revision of Bloom's Taxonomy of Educational Objectives*, Boston MA: Allyn & Bacon (Pearson Education Group).

Andersson, C., Törnberg, A. & Törnberg, P. (2014) Societal systems – Complex or worse?, *Futures*, vol. 63, pp. 145-157.

Aronson, J.L., Harré, R. & Way, E.C. (1995) *Realism Rescued: How Scientific Progress is Possible*. Chicago: Open Court.

Assen, S. van, Boomen, T. van den, Broekman, M., Eijck, G. van, Frijters, E., Klums, M., Lofvers, W., Naafs, S., Spaandonk, T. van, Steketee, A. & Ziegler, F. (2017) *Urban Challenges, Resilient Solutions. Design Thinking for the Future of Urban Regions*, Amsterdam etc: Trancity Valiz with Lectorate Future Urban Regions and the Academies of Architecture in Amsterdam, Rotterdam, Groningen, Arnhem, Tilburg and Maastricht.

Asafu-Adjaye, J., Blomqvist, L., Brook, B., DeFries, R., Ellis, E., Foreman, C., Keith, D., Lewis, M., Lynas, M., Nordhaus, T., Pielke, R., Pritzker, R., Roy, J., Sagoff, M., Shellenberger, M., Stone, R. & Teague, P. (2015) *An Ecomodernist Manifesto*, http://www.ecomodernism.org/

Bailer-Jones, D.M. (2003) When scientific models represent, *International Studies in the Philosophy of Science*, vol. 17, no. 1, pp. 59-74.

Bais, S. (2009) 'Popper voorbij?' In (ibid.) *Keerpunten. Momenten van waarheid in de natuurwetenschappen*, Amsterdam: Amsterdam University Press, pp. 160-165.

Banathy, B. & Jenks, L. (1993) The Transformation of Education: By Design, *International Journal of Educational Research*, vol.19, no. 2, pp.105-115.

Banathy, B. (1999) Systems Thinking in Higher Education, *Systems Research and Behavioral Science*, vol. 16, no. 2, pp. 133-145.

Banathy, B. (2001) We Enter the Twenty-First Century with Schooling Designed in the Nineteenth, *Systems Research and Behavioral Science*, vol. 16, no. 2, pp. 287-290.

Bankes, S.C., Lempert, R.L. & Popper, S.W. (2001) Computer-assisted reasoning, *Computing in Science and Engineering*, vol. 3, pp. 71-77.

Barbieri, M. (2007) Has Biosemiotics Come of Age? In: ibid. (ed.), *Introduction to Biosemiotics. The New Biological Synthesis*, Dordrecht: Springer, pp. 101-113.

Barbieri, M. (2008) Biosemiotics: a new understanding of life, *Naturwisschenschaften*, vol. 95, pp. 577-599.

Barnosky, A., Hadly, E., Bascompte, J., Berlow, E., Brown, J., Fortelius, M., Getz, W., Harte, J., Hastings, A., Marquet, P., Martinez, N., Mooers, A., Roopnarine, P., Vermeij, G., Williams, J., Gillespie, R., Kitzes, J., Marshall, C., Matzke, N., Mindell, D., Revilla, E. & Smith, A. (2012) Approaching a State Shift in Earth's Biosphere, *Nature*, vol. 486. 7 June, pp. 52-58. DOI:10.1038/nature11018

Barth, M. & Michelson, G. (2013) Learning for change: an educational contribution for sustainable science, *Sustainable Science*, No. 8, pp. 103-119.

Barth, M. (2015) *Implementing Sustainability in Higher Education. Learning in an age of transformation*, London/New York: Routledge.

Batty, M. & Xie, Y. (1997) Possible urban automata, *Environment and Planning B – Planning and Design*, vol. 24, pp. 275-292.

Baynes, K., Bohman, J. & McCarthy, Th. (1987) (eds), *After Philosophy*, Cambridge, MA: MIT Press.

Beck, U. (1986) *Risikogesellschaft. Auf dem Weg in eine andere Moderne*, Frankfurt am Main: Suhrkamp Verlag.

Beck, U., Giddens, A., & Lash, S. (1994) *Reflexive Modernization*, Cambridge: Polity Press.

Beddoe, R., Costanza, R., Farley, J., Garza, E., Kent, J., Kubiszewski, I., Martinez, L., McCowen, T., Murphy, K., Myers, N., et al. (2009) Overcoming systemic roadblocks to sustainability: The evolutionary redesign of worldviews, institutions, and technologies, *Proceedings of the National Academy of Sciences*, vol. 106, no. 8, pp. 2483-2489.

Beisbart, C. (2012) How can computer simulations produce new knowledge? *European Journal for Philosophy of Science*, vol. 2, pp. 395-434.

Bekkem, H. van & Lotz, B (2013) Duurzame landbouw: Met of zonder gentech? *GMP*, Vraaggesprek tussen Herman van Bekkem, campagneleider bij Greenpeace, en onderzoeker Bert Lotz van de Wageningen UR, vol. 12:10-11.

Bell, D. (1976) *The coming of the post-industrial society, a venture in social forecasting*, New York: Basic Books.

Berger, P. & Luckmann, P. (1967) *The social construction of reality: A treatise on the sociology of knowledge*, New York: Doubleday & Company.

Bergmann, M., Brohmann, B., Hoffmann, E., Loibl, M. & Rehaag, R. (2005) *Quality criteria of transdisciplinary research. A guide for the formative evaluation of research projects*, Frankfurt am Main (ISOE Studientexte, 13). www.isoe.de/ftp/evalunet_guide.pdf

Bernstein, R.J. (1983) *Beyond Objectivism and Relativism: Science, Hermeneutics, and Praxis*, Philadelphia: University of Pennsylvania Press.

Berry, D. (2011) The computational turn: thinking about the digital humanities, *Culture Machine*, pp. 1465-4121.

Berry, D. (ed.) (2012) *Understanding Digital Humanities*, London: Palgrave/MacMillan.

Bertalanffy, L. von (1969) *General System Theory: Foundations, Development, Applications*, New York: George Braziller Inc.

Bhaskar, R. (1975) *A Realist Theory of Science*, Sussex: The Harvester Press.

Bhaskar, R. (1986) *Scientific Realism and Human Emancipation*, London: Verso.

Bhaskar, R. (1989) *Reclaiming Reality: A Critical Introduction to Contemporary Philosophy*, London: Verso.

Bhaskar, R. (1991) *Philosophy and the Eclipse of Reason: Towards a Metacritique of the Philosophical Tradition – Volume I: Philosophy and the Idea of Freedom*, Oxford/Cambridge: Blackwell.

Boer, Th. de (1980) *Grondslagen van een kritische psychologie*, Baarn: Ambo.

Boer, Th. de (1988) *Filosofische grondslagen van de mens- en cultuurwetenschappen*, Meppel/Amsterdam: Boom.

Boer, Y. de, Gier, A. de, Verschuur, M. & Wit, B. de (2006) *Building Bridges. Researchers on their experiences with interdisciplinary research in the Netherlands*, RMNO, KNAW, NOW, COS, pp. 12-15.

Bod, R. (2010) *De vergeten wetenschappen. Een geschiedenis van de humaniora*. Amsterdam: Bert Bakker.

Bos, B. & Grin, J. (2008) "Doing" Reflexive Modernization in Pig Husbandry – The Hard Work of Changing the Course of a River, *Science, Technology & Human Values*, Vol. 33, No. 4, pp. 480-507.

Bourdieu, P. (1975) The Specificity of the Scientific Field and the Social Conditions of the Progress of Reason, *Social Scientific Information*, vol. 14, no. 6, pp. 19-47.

Bourdieu, P. (1989) *Opstellen over smaak, habitus en het veldbegrip*, Amsterdam: Van Gennep.

Bourdieu, P. & Passeron, J.C. (1990) *Reproduction in Education, Society and Culture*, London: Sage Publications.

Bradie, M.A. (2006) Scientific Progress, in: Sarkar, J. & J. Pfeifer (eds) *The Philosophy of Science – An Encyclopedia of Philosophy*, New York: Routledge, pp. 749-753.

Braseby, A. (2015) The Flipped Classroom, *Idea paper 57*, Retrieved 25 May 2015 from http://ideaedu.org/research-and-papers/idea-papers/idea-paper-no-57/

Braungart, M. & McDonough, W. (2002) *Cradle to Cradle: Remaking the Way We Make Things*, North Point Press.

Braungart, M., McDonough, W. & Bollinger, A. (2007) Cradle-to-cradle design: creating healthy emissions – a strategy for eco-effective product and system design, *Journal of Cleaner Production*, vol.15, pp. 1337-1348.

Brigandt, I. (2016) Do we need a 'theory' of development? *Biological Philosophy*, vol. 31, pp. 603-617.

Brown, L. (2001) *Eco-Economy – Building an Economy for the Earth*, Earth Policy Institute. London: Earthscan.

Brown, J.R. (2004) Why thought experiments transcend empiricism, in: C. Hitchcock (ed.), *Contemporary Debates in Philosophy of Science,* Oxford: Blackwell, pp. 23-43.

Brown, Valerie., Harris, John., Russell, Jacqueline. (2010) *Tackling Wicked Problems Through the Transdisciplinary Imagination,* London/Washington D.C.: Earth Scan.

Bruchatz, C., Fischer, R., Herzer, P., Meyer, M. & Stelzer, J. (2018) *Applying Design Thinking – A Workbook on Methods for Teachers in Higher Education and Researchers,* Center for Synergy Enhancement in commission of Erasmus+ Programme of the European Union.

Bryman, A. (2004) *Social Research Methods,* Oxford: Oxford University Press.

Bryson, B. (2003) *A Short Theory of Nearly Everything,* USA: Broadway Books.

Buclet, N. & Lazarevic, D. (2015) Principles for Sustainability: the need to shift to a sustainable conventional regime, *Environmental Developmental Sustainability,* vol. 17, pp. 83-100.

Byrne, D. (2002) *Interpreting quantitative data,* London: Sage.

Byrne, D. (2005) Complexity, Configurations and Cases, *Theory, Culture & Society,* vol. 22, no. 5, pp. 95-111.

Byrne, D. (2014) *Complexity and the Social Sciences,* London/New York: Routledge.

Byrne, D. & Callaghan, G. (2013) *Complexity Theory and the Social Sciences: The state of the art,* London: Routledge.

Canton, E., Lanser, D., Noailly, J., Rensman M. & Ven, J. van de (2005) *Crossing borders: when science meets industry* (2005) Netherlands Bureau for Economic Policy Analysis (CPB), document 98, 5 October 2005.

Capra, F. (2001) Life and Leaderschip in Organizations, in: (ibid.)*The Hidden Connections: A Science for Sustainable Living,* London: Harper Collins.

Carnap, R. (1932-3) Über Protokollsätze, *Erkenntnis,* vol. 3, pp. 215-228.

Cartledge, K., Dürrwächter, C., Jimenez, V.H., Winder, N.P. (2009) Making sure you solve the right problem, *Ecological Sociology,* vol. 14 : r3.

Castree, N. (2005) *Nature,* London: Routledge.

Castree, N., Adams, W.M., Barry, J., Brockington, D., Büscher, B., Corbera, E., Demeritt, D., Duffy, R., Felt, U., Neves, K., et al. (2014) Changing the intellectual climate, *Nature climate change,* vol. 4, no.9, pp. 763-768.

Chalmers, A. (1999) (3r ed.) *What is this thing called science?* St. Lucia: Queensland University Press, co-published by Indianapolis/Cambridge: Hackett Publishing Company.

Cilliers, P. (2001) *Complexity and Postmodernism. Understanding Complex Systems,* London: Routledge.

Cilliers, P. (2000) Knowledge, Complexity and Understanding, *Emergence,* vol. 2, no. 4, pp. 7-13.

Cilliers, P. (2001) Boundaries, hierarchies and networks in complex systems, *International Journal of Innovation Management,* vol. 5, no. 2, pp. 135-147.

Cilliers, P. (2002) Why We Cannot Know Complex Things Completely, *Emergence,* vol. 4, no.1-2, pp. 77-84.

Clayton, A., Radcliffe, N. (1996) *Sustainability – A Systems Approach.* London: Earthscan Publications.

Coenen, H. (1987) *Handelingsonderzoek als exemplarisch leren – een bijdrage aan de fundering van de methodologie van handelingsonderzoek*, Groningen: Konstapel.

Coenen H. (2001) Handelingsonderzoek – De verhouding tussen onderzoeker en onderzochte, *Tijdschrift voor Arbeid en Participatie*, jrgng. 23, no. 1, pp. 63-74.

Collier, A. (1994) *Critical Realism*, London/New York: Verso.

Commissie Toekomstbestendig Hoger Onderwijs (2010) *Differentiëren in drievoud omwille van kwaliteit en verscheidenheid in het hoger onderwijs*, Den Haag: Advies van de Commissie Toekomstbestendig Hoger Onderwijs, april 2010.

Crutzen, P. & Stoermer, E. (2000) The Anthropocene, Global Change, *IGBP Newsletter,*vol. 41, pp. 17-18.

d'Ancona, M. (2017) *Post-Truth. The New War on Truth and How To Fight Back,* Ebury Publishing.

Dahms, H.J. (1992) Positivismus und Pragmatismus, in: D. Bell & W. Vossenkuhl (Hrsg.), *Wissenschaft und Subjektivität*, Berlin: Akademie Verlag, pp. 239-257.

Daly, H. (1996) *Beyond Growth – The Economics of Sustainable Development.* Boston: Beacon Press.

Darwin, Ch. (1859) *On the Origin of Species by Means of Natural Selection, or the Preservation of Favoured Races in the Struggle for Life* (1st ed.). London: John Murray.

Davis, E. (2017) *Post-Truth. Why We Have Reached Peak Bullshit and What We Can Do About It,* London: Little Brown.

DeTombe, D.J. (1994) *Defining Complex Interdisciplinary Societal Problems. A Theoretical Study for Constructing a Cooperative Problem Analyzing Method: The Method COMPRAM*, Amsterdam: Thesis Publishers.

DeTombe, D.J. (2015) *Handling* Societal Complexity Amsterdam: Thesis Publishers.

Dehue, T., *De regels van het vak. Nederlandse psychologen en hun methodologie 1900-1985*, Amsterdam: Van Gennep, 1990.

Denzin, N. & Lincoln, Y. (1994) *Handbook of Qualitative Research*, Thousand Oakes: SAGE.

Diamond, J. (1999) *Guns, Germs, and Steel. The Fates of Human Societies*, New York/ London: Norton & Company.

Diamond, J. (2004) *Collapse, How Societies Choose to Fail or Succeed,* New York: Viking.

Dijstelbloem, H., Huisman, F., Miedema, F., Ravetz, J., & Mijnhardt, W. (2013) *Position Paper Science in Transition*, http://www.scienceintransition.nl.

Dilthey, W. (1881) *Einleitung in die Geisteswissenschaften. Versuch einer Grundlegung für das Studium der Gesellschaft und der Geschichte*, Leipzig: Tuebner.

Dyson, F. (2013) Hoe onverenigbare wereldbeelden toch naast elkaar kunnen bestaan, in: Brockman, J. (ed.) *Dit verklaart alles.* Amsterdam: Maven Publishing, pp. 99-100.

Educause (2012) *Seven things you should know about flipped classrooms*, Educause Learning Initiative, Retrieved 5 July 2015 from https://net.educause.edu/ir/ library/pdf/eli7081.pdf

Engelen, E. & Thieme, M. (2016) *De kanarie in de kolenmijn*, Amsterdam: Prometheus.

Ericksen, P.J. (2007) Conceptualizing food systems for global environmental change research. *Global Environmental Change*, vol. 18, no. 1, pp. 234-245.

European Commission (2010) *Assessing Europe's University-Based Research*, Brussels: European Commission.

FAO (2004) *The State of Food and Agriculture. Agricultural Biotechnology: Meeting the needs of the poor?*, Food & Agricultural Organization of the United Nations, Rome: FAO Agricultural Series, No. 35.

Fay, B. (1996) *Contemporary Philosophy of Social Science*, Oxford: Blackwell.

Frigg, R., & Reiss, J. (2009) The philosophy of simulation: Hot new issues or some old stew? *Synthese*, vol. 169, pp. 593-613.

Feyerabend, P. (1975) *Against Method*, London: New Left Books.

Fletcher, J. (1966) *Situation Ethics*, Philidelphia: Westminster.

Flyvbjerg, B. (2001) *Making Social Science Matter: Why Social Inquiry Fails and How it can Succeed Again*, Cambridge: Cambridge University Press.

Folke, C., Carpenter, S.R., Walker, B., Scheffer, M., Chapin, T., Rockstrom, J. (2010) Resilience thinking: integrating resilience, adaptability and transformability, *Ecology & Society*, vol. 15, no. 4.

Foucault, M. (1976) *De orde van het vertoog*, Meppel: Boom.

Foucault, M. (1979) *Discipline and Punish*, New York: Vintage.

Foucault, M. (1980) *Power/Knowledge*, Brighton: Harvester.

Fresco, L. (2003) Which Road Do We Take? Harnessing Genetic Resources and Making Use of Life Sciences, a New Contract for Sustainable Agriculture, paper for the EU Discussion Forum *Towards Sustainable Agriculture for Developing Countries: Options from Life Sciences and Biotechnologies*, Brussels, 30-31 January 2003, Rome: Food and Agriculture Organization of the United Nations.

Fresco, L. (2006) Sustainable Agro-Food Chains, Challenges for Research and Development, Rome: Food and Agriculture Organization of the United Nations. Ruben, R., Slingerland, M. & Nijhoff, H. (eds) (2006) *Sustainable Agro-Food Chains and Networks for Development*, Dordrecht: Springer, pp. 205-208.

Fresco, L. (2009) Challenges for food system adaptation today and tomorrow, *Environmental Science & Policy*, vol. 12, no. 4, pp. 378-385.

Fresco, L. (2012) *Hamburgers in het paradijs – Voedselschaarste in tijden van overvloed*, Amsterdam: Bert Bakker.

Fresco, L. (2013) The GMO Stalemate in Europe, *Science*, vol. 339, no. 6122: 883.

Fresco, L. (2014) Some thoughts about the future of food and agriculture, *South African Journal of Science*, vol.110, no.5-6, March 2014.

Fresco, L. (2015) The new green revolution: Bridging the gap between science and society, *Current Science*, vol. 109, no. 3, 10 August 2015, pp. 430-438.

Funtowicz, S. & Ravetz, J.R (1994) Emergent complex systems, *Futures*, vol. 26, no. 6, pp. 568-582.

Gadamer, H.-G.(1960) *Wahrheit und Methode*, Tübingen: Mohr.

Geels, F.W. (2002) Technological transitions as evolutionary reconfiguration processes: a multi-level perspective and a case-study, *Research Policy*, vol. 31, no. 8-9, pp. 1257-1274.

Gibbons, M., Limoges, C., Nowotny, H., Schwarzman, S., Scott, P. & Trow, M. (1994) *The New Production of Knowledge: The Dynamics of Science and Research in Contemporary Societies*, London: Sage.

Giddens, A. (1976) *New Rules of Sociological Method*, London: Hutchinson.

Giddens, A. (1985) Structuratietheorie en empirisch onderzoek, Gastcollege uitgesproken te Wageningen op 27 april 1984, in: Q. Munters, E. Meijer, H. Mommaas, H. van der Poel, R. Rosendal & G. Spaargaren, *Een kennismaking met de structuratietheorie*, Wageningen: Landbouwhogeschool Wageningen, pp. 27-56.

Giddens, A. (1990) *The Consequences of Modernity*, Cambridge: Polity Press.

Giddens, A. (1991) *Modernity and Self-identity*, Cambridge: Polity Press.

Giddens, A. (1992) *The Transformation of Intimacy: Sexuality, Love, and Eroticism in Modern Societies,* Cambridge: Cambridge University Press.

Giere, R.N. (2004) How models are used to represent reality, *Philosophy of Science*, vol. 71, pp. 742-752.

Giere, R.N. (2006) The role of agency in distributed cognitive systems, *Philosophy of Science*, vol. 73, no. 5, pp. 710-719.

Giere, R.N. (2010) An agent-based conception of models and scientific representation. *Synthese*, vol. 172, no. 2, pp. 269-281.

Gilbert, N., & Troitzsch, K.G. (1999) *Simulation for the social scientist,* Buckingham: Mcgraw-Hill Professional.

Godeman, J. (2008) Knowledge integration: A key challenge for transdisciplinary cooperation, *Environmental Education Research,* Vol. 14, No. 16, pp. 625-641. DOI: 10.1080/13504620802469188.

Godfrey-Smith, P. (2006) The strategy of model-based science, *Biology and Philosophy*, vol. 21, no. 5, pp. 725-740.

Gore, A. (2013) *The Future. Six Drivers of Global Change*, New York: Random House.

Gouvea, J. & Passmore, C. (2017) 'Models of' versus 'Models for' – Toward an Agent-Based Conception of Modeling in the Science Classroom, *Science & Education*, Vol. 26, pp. 49-63.

Greenpeace (2015) *Twenty Years of Failure. Why GM crops have failed to deliver on their promises*, Greenpeace, November 2015. www.greenpeace.nl/Global/nederland/report/2015/Landbouw/20_years_GMO_failure.pdf Accessed 19 July 2016.

Greenwood, D. & Levin, M. (1998) *Introduction to Action Research, Social Research for Social Change*, Thousand Oakes: SAGE.

Gregory, W. (1993) Designing Educational Systems: A Critical Systems Approach, *Systems Practice*, vol. 6, no. 2, pp. 199-209.

Grin, J. (2005) Reflexive modernisation as a governance issue, or designing and shaping *re*-structuration, in: Voß, J-P., Bauknecht, D., Kemp, R. (eds) (2005) *Reflexive Governance for Sustainable Development*, Cheltenham: Edward Elgar.

Grin, J. (2012) Changing Government, Kitchens, Supermarkets, Firms and Farms, in: Spaargaren, G., Oosterveer, P. & Loeber, A. (2012) *Food Practices in Transition. Changing Food Consumption, Retail and Production in the Age of Reflexive Modernity*, New York/London: Routledge, pp. 35-59.

Grin, J., Felix, F. & Bos, B. (2004) Practices for reflexive design: Lessons from a Dutch programme on sustainable agriculture, *International Journal Foresight and Innovation Policy*, vol. 1, no. 1/2, pp. 126-149.

Grin, J., Rotmans, J. & Schot, J. (2010) *Transitions to Sustainable Development. New Directions in the Study of Long Term Transformative Change*, London/New York: Routledge.

Grin, J. & Weterings, R. (2005) *Reflexive monitoring of system innovative projects: strategic nature and relevant competencies*, Paper presented at the 6th Open Meeting of the Human Dimensions of the Global Environmental Change Research Community at the University of Bonn, Germany, October.

Grobstein, P. (2007) From Complexity to Emergence and Beyond, *Soudings*, vol. XC, no. 1-2, pp. 9-31.

Grüne-Yanoff, T., & Weirich, P. (2010) The philosophy and epistemology of simulation. A review. *Simulation & Gaming*, vol. 41, pp. 20-50.

Guignon, Ch. (1983) Pragmatism or Hermeneutics? Epistemology after Foundationalism, in: D.R., Hiley, J.F. Bohman, & R. Shustermans (eds), *The Interpretative Turn. Philosophy, Science, Culture*, New York: Cornell University Press, pp. 81-101.

Haasnoot, M., Kwakkel, J.H., Walker, W.E., ter Maat, J. (2013) Dynamic adaptive policy pathways: a method for crafting robust decisions for a deeply uncertain world, *Global environmental change*, vol. 23, no. 2, pp. 485-498.

Habermas, J.(1970) *Zur Logik der Sozialwissenschaften*, Frankfurt am Main: Suhrkamp Verlag.

Habermas, J. (1973) *Legitimationsprobleme im Spätkapitalismus*, Frankfurt am Main: Suhrkamp Verlag.

Habermas, J. (1981a) *Theorie des kommunikativen Handelns I & II*, Frankfurt am Main: Suhrkamp Verlag. (In English: *The Theory of Communivative Action*, Boston: Beacon Press, 1984-1987, Translation by T. McCarthy).

Habermas, J. (1981b) Die Moderne – Ein unvollendetes Projekt, in: idem, *Kleine Politische Schriften I-IV*, Frankfurt am Main: Suhrkamp Verlag, pp. 444-464.

Habermas, J. (1996) *Die Einbeziehung des Anderen. Studien zur politischen Theorie*, Frankfurt am Main: Suhrkamp Verlag. (In English: *The Inclusion of the Other. Studies in Political Theory*, Cambridge MA/London: MIT Press, 1998.)

Hajer, M. (2017) *De macht van verbeelding*, Utrecht: Universiteit Utrecht.

Harari, Y. (2017) *Homo Deus. A Brief History of Tomorrow*, London: Vintage.

Hardin, G. (1968) The Tragedy of the Commons, *Science* (New Series), vol. 162, no. 3859, pp. 1243-1248.

Haring, M. & Van Bekkem, H. (2013) *Who owns our food? Over plantenveredeling, patentering, globarisering en GM planten*, FNWI Collegetour in samenwerking met Faculteit der Natuur-wetenschappen. http://www.spui25.nl/programma/item/10.04.13---who-owns-your-food.html

Hart, H. 't, Dijk, J. van, Goede, M. de, Jansen, W. & Teunissen, J. (2001) *Onderzoeksmethoden*, Amsterdam: Boom.

Hartmann, S. (1996) The world as a process: Simulation in the natural and social sciences. In R. Hegselmann, U. Müller, & K.G. Troitzsch (eds) *Modelling and simulation in the social sciences from the philosophy of science point of view*, Dordrecht: Kluwer, pp. 77-100.

Harvey, D.L. & Reed, M. (1996) Social science as the study of complex systems, *Chaos theory in the social sciences*, pp. 295-324.

Helbing, D. (2013) Globally networked risks and how to respond, *Nature*, 497 (7447), pp. 51-59.

Hempel, C.G. (1965) Aspects of Scientific Explanation, in: idem., *Aspects of Scientific Explanation, and Other Essays in the Philosophy of Science*, New York/London: The Free Press/Collier-MacMillan Limited, pp. 331-496.

Hirsch Hadorn, G., Bradley, D., Pohl, C., Rist, S., Wiesmann, U. (2006) Implications of transdisciplinarity for sustainability research, *Ecological Ecomonics*, vol. 6, pp. 119-128.

Hirsch Hadorn, G., Hoffmann-Riem, H., Biber-Klemm, S., Grossenbacher-Mansuy, W., Joye, D., Pohl, C., Wiesmann, U. & Zemp, E. (eds) (2008) *Handbook of Transdisciplinary Research*, Dordrecht: Springer.

Hirsch Hadorn, G., Pohl, D. & Bammer, G. (2017) Solving Problems through Transdisciplinary Research, in: Frodeman, R., Thompson Klein, J. & Pacheco, R. (eds) *The Oxford Handbook of Interdisciplinarity* (2nd edition), Oxford: Oxford University Press, pp. 431-450.

Hoffmeyer, J. (2008) *Biosemiotics. An Examaniation into the Signs of Life and the Life of Signs*, Chicago: University of Scranton Press.

Hofstadter, D. (1985) *Metamagical Themas*: Questing for the Essence of Mind and Pattern, New York: Basic Books.

Hollis, M. (1994) *The Philosophy of Social Science*, Cambridge: Cambridge University Press.

Homer-Dixon, T. (2011) Complexity Science. Shifting the trajectory of civilisation, *Oxford Leadership Journal*, Vol. 2, No. 1, pp. 1-15.

Huxley, J. (1942) *Evolution: The Modern Synthesis*, London: Allen & Unwin (2nd ed. 1963).

International Science Cooperation to Meet Global 'Grand Challenges' (2010) *Earth System Science for Global Sustainability: The Grand Challenges*, Paris: International Social Science Council.

Ison, R. (1999) Applying Systems Thinking to Higher Education, *Systems Research and Behavioral Science*, vol. 16, pp. 107-112.

Ison, R. (2001) Systems practice at the United Kingdom's Open University, in: Wilby, J. & Ragsdell, G. (eds) *Understanding Complexity*, Kluwer Academic, pp. 45-54.

Jasanoff, S. (2004) *States of Knowledge – The co-production of science and social order*, London/New York: Routledge.

Jasanoff, S. (2014) *Science and Public Reason*, London/New York: Routledge.

Jasanoff, S. (2017) Future Imperfect: Science, Technology, and the Imaginations of Modernity, in: Jasanoff, S. & Sang-Hyun, K. (eds) (2017) *Dreamscapes of Modernity – Sociotechnical Imaginaries and the Fabrication of Power*, Chicago: University of Chicago Press, pp. 1-33.

Kant, I. (1781, 1996 transl. by W.S. Pluhar) *Critique of Pure Reason*, Indianapolis/Cambridge: Hackett Publishing.

Kant, I. (1784) *Beantwortung der Frage:* Was ist Aufklärung? *Berlinische Monatsschrift*, December 1984.

Kennedy, E., Gladek, E. & Roemer, G. (2018) *Using Systems Thinking to Transform Society*, Amsterdam: Metabolic, in cooperation with Milieusamenwerking en Afvalverwerking Regio Nijmegen, Gemeente Nijmegen, ARN BV en DAR NV commissioned by WWF.

Kerkhoff, L. van (2014) Developing integrative research for sustainability science through a complexity principles-based approach, *Sustainability Science*, vol. 9, pp. 143-155.

Kjærnes, U. & Torjussen, H., (2012) Beyond the Industrial Paradigm?, in: Spaargaren, G., Oosterveer, P. & Loeber, A. (2012) *Food Practices in Transition. Changing Food Consumption, Retail and Production in the Age of Reflexive Modernity*, New York/London: Routledge, pp. 86-106.

Klein, N. (2017) *No is Not Enough*, London: Allen Lane.

Kloppenburg, J., Hendrickson, J., & Stevenson, G.W. (1996) Coming in to the foodshed, *Agriculture and Human Values*, vol. 13, no. 3, pp. 33-42.

Klukhuhn, A. (2008) *Alle mensen heten Janus. Het verbond tussen filosofie, wetenschap, kunst en godsdienst*, Amsterdam: Bert Bakker.

Knuuttila, T. (2005) Models, representation, and mediation, *Philosophy of Science*, vol. 72, no. 5, pp. 1260-1271.

Knuuttila, T. (2011) Modelling and representing: An artefactual approach to model-based representation, *Studies in History and Philosophy of Science Part A*, vol. 42, no. 2, pp. 262-271.

Koningsveld, H. (1987) *Het verschijnsel wetenschap*, Amsterdam/Meppel: Boom.

Krohn, W., & Daele, W. van den, (1998) Experimental implementation as a linking mechanism in the process of innovation, *Research Policy*, vol. 27, pp. 853-868.

Kuhn, T.S. (1962) *The Structure of Scientific Revolutions*, Chicago: University of Chicago Press.

Kuhn, T.S. (1970) Logic of Discovery or Psychology of Research?, in: Lakatos et al., pp. 1-24.

Kunneman, H. (1986) *De waarheidstrechter – een communicatietheoretisch perspektief op wetenschap en samenleving*, Meppel: Boom.

Kunneman, H. (2005) *Voorbij het dikke-ik. Bouwstenen voor een kritisch humanisme*, Amsterdam: SWP.

Kwa, C. (2005) *De ontdekking van het weten. Een andere geschiedenis van de wetenschap*, Amsterdam: Boom.

Kwa, C. (2011) *Styles of knowing; A new history of science from ancient times to the present*, Pittsburgh PA: University of Pittsburgh Press.

Kwa, C. (2014) *Kernthema's in de wetenschapsfilosofie*, Den Haag: Boom Lemma.

Lakatos, I. (1978) *The Methodology of Scientific Research Programmes*, Cambridge: Cambridge University Press.

Lakatos, I. & Musgrave, A. (eds) (1970) *Criticism and the Growth of Knowledge*, Cambridge: Cambridge University Press, pp. 91-196.

Lane, D. (2011) Complexity and innovation dynamics, in: Antonelli, C. (ed.), *Handbook on the Economic Complexity of Technological Change*, Cheltenham: Edward Elgar Publishing, pp. 63-80.

Lane, D.A., van der Leeuw, S., Sigaloff, C., Addarii, F. (2011) Innovation, sustainability and ict, *Procedia Computer Science*, vol. 7, pp. 83-87.

Latour, B. (1993) *We have never been modern*, Cambridge MA: Harvard University Press.

Latour, B. (1999) *Pandora's Hope – Essays on the Reality of Science Studies*, Cambridge MA/London: Harvard University Press.

Latour, B. (2003) Is Re-modernization Occurring – And If So, How to Prove It? A Commentary on Ulrich Beck, *Theory, Culture & Society*, vol. 20, no. 2, pp. 35-48.

Latour, B. (2017) *Facing Gaia. Eight lectures on the new climate regime*, Cambridge: Polity Press. Content: https://www.amazon.com/Facing-Gaia-Lectures-Climatic-Regime/dp/0745684335

Laursen, K. & Salter, A. (2004) Searching high and low: what types of firms use universities as a source of innovation? *Research Policy*, vol. 33, no. 8: 1201-1215.

Leach, M., Scoones, I., Stirling, A. (2010) *Dynamic sustainabilities: technology, environment, social justice*, Earthscan.

Leezenberg, M. & Vries, G. de (2005) *Wetenschapsfilosofie voor geesteswetenschappen*, Amsterdam: Amsterdam University Press.

Lemaire, T. (2010) *De val van Prometheus – Over de keerzijden van de vooruitgang*, Amsterdam: Ambo.

Levine & Côté (2002) *Identity, Formation, Agency & Culture. A Social Psychological Synthesis*, Mahwah/London: Lawrence Erlbaum Associates Publishers.

Loorbach, D. (2010) Transition management for sustainable development: a prescriptive, complexity-based governance framework, *Governance*, vol 23, no.1, pp. 161-183.

Love, A.C. (2014) The erotetic organization of developmental biology, in: Minelli, A. & Pradeu, T. (eds) (2014) *Towards a theory of development*, Oxford: Oxford University Press.

Lyotard, J.-F. (1984) *The Postmodern Condition: A Report on Knowledge*, Manchester: Manchester University Press.

Maat, H. (2011) Voedselzekerheid en de onzekerheid van wetenschap, in: Dijstelbloem, H. & R. Hagendijk (red.) *Onzekerheid troef? Het betwiste gezag van wetenschap*, Amsterdam: Van Gennep, pp. 157-174.

Maiteny, P. & Ison, R. (2000) Appreciating Systems: Critical Reflections on the Changing Nature of Systems as a Discipline in a Systems-Learning Society, *Journal of Systemic Practice and Action Research*, vol. 13, no. 4, pp. 559-586.

Manuel-Navarrete, D., Buzinde, C. (2010) Socio-ecological agency: From "human exceptionalism" to coping with "exceptional" global environmental change, in: Redclift, M., Woodgate, G. (eds), *The International Handbook of Environmental Sociology*, Cheltenham: Edward Elgar, pp. 136-149.

Marsden, T. (2012) Food Systems Under Pressure, in: Spaargaren, G., Oosterveer, P. & Loeber, A. (2012) *Food Practices in Transition. Changing Food Consumption, Retail and Production in the Age of Reflexive Modernity*, New York/London: Routledge, pp. 291-311.

Martinuzzi, A. & Sedlaçko, M. (2016) *Knowledge Brokerage for Sustainable Development*, Saltaire: Greenleaf Publishing.

Mauser, W., Klepper, G., Rice, M., Schmalzbauer, B.S., Hackmann, H., Leemans, R., Moore, H. (2013) Transdisciplinary Global Change Research: the co-creation of knowledge for sustainability, *Current Opinion on Environmental Sustainability*, vol. 5, no. 3-4, pp. 420-431.

Maxwell, N. (2008) Are Philosophers Responsible for Global Warming? *Philosophy Now*, January/February 2008, pp. 12-13.

McCarthy, D., Crandall, D., Whitelaw, G., General, Z. & Tsuji, L. (2011) A Critical Systems Approach to Social Learning: Building Adaptive Capacity in Social, Ecological, Epistemological (SEE) Systems, *Ecology and Society*, vol. 16, no. 3, Article 18.

McCarthy, Th. (1978) *The Critical Theory of Jürgen Habermas*, London: Hutchinson.

McCarthy, Th. (1991) *Ideals and Illusions: On Reconstruction and Deconstruction in Contemporary Critical Theory*, Cambridge MA/London: MIT Press.

McCracken, G. (2006) *Flock and Flow. Predicting and Managing Change in a Dynamic Marketplace*, Bloomington & Indianapolis: University of Indiana.

McMurtry, A. & Dellner, J. (2014) Relationalism: An Interdisciplinary Epistemology. Or, why our knowledge is more like a coral reef than fish scales, *Integrative Pathways* (Newsletter of the Association for Interdisciplinary Studies), October 2014, vol. 36, no. 3, pp. 6-12.

Meadows, D.H., Meadows, D.L., Randers, J., & Behrens III, W.W. (1972) *The Limits to Growth*, A Report for the Club of Rome's Project on the Predicament of Mankind, New York: Universe Books.

Meadows, D.H., Meadows, D.L. & Randers, J. (1992) *Beyond the Limits – Global Collapse or a Sustainable Future*, London: Earthscan.

Meadows, D.H. (1999) *Leverage Points: Places to Intervene in a System*, Hartland: Sustainability Institute.

Meadows, D. (2008) *Thinking in Systems. A primer*, Vermont: Chelsea Green Publishing.

Melchett, P. (2012) The pro-GM lobby's seven sins against science, www.soilassociation.org/motherearth/viewarticle/articleid/4752/the-pro-gm-lobbys-seven-sins-against-science, retrieved 21 June 2014.

Menken, S. & Keestra, M. (eds) (2016) *An Introduction to Interdisciplinary Research*, Amsterdam: Amsterdam Univerity Press.

Merriënboer, J.J.G. van, Clark, R.E. & de Croock, M. (2002) Blueprints for complex learning: The 4C/ID-model. *Educational Technology Research and Development*, 50, pp. 39-64.

Merriënboer, J.J.G. van & Sluijsmans, D.M.A. (2009) Toward a Synthesis of Cognitive Load Theory, Four-Component Instructional Design, and Self-Directed Learning. *Educational Psychology Review*, 21, pp. 55-66.

Midgley, G. (2000) *Systemic Intervention: Philosophy, Methodology, and Practice*, New York etc: Kluwer Academic/Plenum Publishers.

Mierlo, B. van, Regeer, B., Amstel, M. van, Arkesteijn, M., Beekman, V., Bunders, J., Cock Buning, T. de, Elzen, B., Hoes, A.C. & Leeuwis, C. (2010) *Reflexive Monitoring in Action. A guide for monitoring system innovation projects*, Amersfoort: AKIMOTO.

Miller, C.A. (2004) Climate Science and the making of a global political order, in: Jasanoff, S. (2004) *States of Knowledge – The co-production of science and social order*, London/New York: Routledge, pp. 46-66.

Miller, C.A. (2017) Globalizing Security: Science and the Transformation of Contemporary Political Imagination, in: Jasanoff, S. & Sang-Hyun, K. (eds) (2017) *Dreamscapes of Modernity – Sociotechnical Imaginaries and the Fabrication of Power*, Chicago: University of Chicago Press, pp. 277-299.

Morrison, M. & Morgan, M.S. (1999) Models as mediating instruments, in: M. Morrison & M.S. Morgan (eds), *Models as mediators. Perspectives on natural and social science*. Cambridge University Press, pp. 10-37.

Morin, E. (2008) *On Complexity*, Cresskill: Hampton Press.

Murdoch, J. (1997) Inhuman/nonhuman/human: actor-network theory and the prospects for a nondualistic and symmetrical perspective on nature and society. *Environment and Planning: Society and Space*, vol. 15, no. 6, pp 731-756.

Nersessian, N.J. (1992) In the theoretician's laboratory: Thought experimenting as mental modeling, *Proceedings of the philosophy of science association*, vol. 2, pp. 291-301, The University of Chicago Press on behalf of the philosophy of Science Association.

Nesse, R. (2013) Natuurlijke selectie is eenvoudig, maar de systemen die erdoor worden gevormd, zijn onvoorstelbaar complex, In: Brockman, J. (ed.) *Dit verklaart alles*, Amsterdam: Maven Publishing, pp. 337-340.

Newell, W.H. (2011) The road from interdisciplinary studies to complexity, *World Futures*, vol. 67, pp. 330-342.

Neurath, O., Carnap, A. & Morris, Ch. (eds) (1952) *International Encyclopedia of Unified Science*, Chicago: University of Chicago Press.

Newell, W.H. (2011) The Road from Interdisciplinary Studies to Complexity, *World Futures*, vol. 67, no. 4-5, pp. 330-342.

Nicolopoulou, K. (2011) Towards a theoretical framework for knowledge transfer in the field of CSR and sustainability, *Equality, Diversity and Inclusion: An International Journal*, vol. 30, no. 6, pp. 524-538.

Nonaka, I. (1994) A dynamic theory of organizational knowledge creation, *Organization Science*, vol. 5, no. 1, pp. 14-37.

Norberg, J. (2017) *Progress – Ten Reasons to Look Forward ro the Future*, London: Oneworld Publications.

Nowotny, H., Scott, P. & Gibbons, M. (2001) *Re-Thinking Science. Knowledge and the Public in an Age of Uncertainty*, Cambridge: Polity Press.

Nowotny, H. (2005) The Increase of Complexity and its Reduction. Emergent Interfaces between the Natural Sciences, Humanities and Social Sciences, *Theory, Culture & Society*, vol. 22, no. 5, pp. 15-31.

Nowotny, H. (2016) *The Cunning of Uncertainty*, Cambridge/Malden: Polity Press.

Nussbaum, M. (1990) *Love's Knowledge. Essays on Philosophy and Literature*, New York/Oxford: Oxford University Press.

O'Brien, K., Reams, J., Caspari, A., Dugmore, Q., Faghihimani, M., Fazey, I., Hackmann, H., Manuel-Navarette, D., Marks, J., Miller, R., Raivio, K., Romero-Lankao, P., Virji, H., Vogel, C., Winiwarter, V. (2013) You say you want a revolution? Transforming education and capacity building in response to global change, *Environmental Science & Policy*, vol. 28, pp. 48-59.

Ochs, E., Jacoby, S., & Gonzales, P. (1994) Interpretive Journeys: How Physicists Talk and Travel through Graphic Space,*Configurations*, vol. 2, no. 1, pp. 151-171.

Ockels, W. (2014) Brief aan de mensheid, *Algemeen Dagblad,* d.d. 17 May 2014.

OECD (2002) *Benchmarking Industry-Science Relationships*, Paris: Organisation for Economic Co-operation and Development.

Olaya, C.(2014) *The Scientist Personality of System Dynamics*, Paper presented at the 32nd International Conference of the System Dynamics Society.

Olsson, L., Jerneck, A., Thoren, H., Persson, J., O'Byrne, D. (2015) Why resilience is unappealing to social science: Theoretical and empirical investigations of the scientific use of resilience, *Science Advances*, 22 May 2015: E1400217, Vol. 1, No. 4, 11 pp. Gedownload van http://advances.sciencemag.org/op 18 February 2016.

Oreskes, N. & Conway, E.M. (2013) The Collapse of Western Civilization. A View from the Future, *Daedalus*, vol. 142, no. 1, pp. 40-58.

Oreskes, N. & Conway, E.M. (2014) *The Collapse of Western Civilization. A View from the Future,* Columbia: Columbia University Press.

Osberg, D., Biesta, G., & Cilliers, P. (2008) From Representation to Emergence: Complexity's challenge to the epistemology of schooling, *Educational Philosophy and Theory*, vol. 40, no. 1, DOI: 10.1111/j.1469-5812.2007.00407.x

Osnos, E. (2017) Survival of the Richest, *The New Yorker*, 30 January 2017.

Parker, B. (2013) Het is gewoon zo? In: Brockman, J. (ed.) *Dit verklaart alles,* Amsterdam: Maven Publishing, pp. 297-300.

Peperstraten, F. van (1993) *Samenleving ter discussie*, Bussum: Coutinho.

Pieters, K. (2010) *Into Complexity. A Pattern-oriented Approach to Stakeholder Communications*, Utrecht: University of Humanistics/NWO.

Pigliucci, M., & Müller, G.B. (2010) *Evolution, the extended synthesis*, Cambridge, MA: MIT Press.

Piketty, T. (2014) *Capital in the Twenty-First Century*, Harvard: Harvard University Press (Translated by Arthur Goldhammer).

Pohl, C. & Hirsch Hadorn, G. (2007) *Principles for Designing Transdisciplinary Research: Proposed by the Swiss Academies of Arts and Sciences,* München: oekom Verlag.

Pohl, C. & Hirsch Hadorn, G. (2008) Methodological Challenges of Transdisciplinary Research, *Natures Sciences Sociétés*, vol. 16, no. 2, pp. 111-121.

Polanyi, M. (1958) *Personal Knowledge: Towards a Post-Critical Philosophy*, Chicago: University of Chicago Press.

Polanyi, M. (1966) *The Tacit Dimension*, Chicago: University of Chicago Press.

Popper, K.R. (1934) *Logik der Forschung*, Wenen: Springer. (In English (1959): *The Logic of Scientific Discovery*, London: Hutchinson).

Popper, K.R. (1957) *The Poverty of Historicism*, London: Routledge & Kegan Paul Ltd,.

Popper, K.R. (1961) *The Logic Of Scientific Discovery.* New York: Science Editions.

Popper, K.R. (1963) *Conjectures and Refutations – The Growth of Scientific Knowledge*, London: Routledge & Kegan Paul Ltd.

Popper, K.R. (1968) Epistemology without a Knowing Subject, in: B. Rootselaar & J. Staal (eds), *Proceedings of the Third International Congress for Logic, Methodology and Philosophy of Science*, Amsterdam: North Holland, pp. 333-373.

Popper, K.R. (1970) Normal Science and its Dangers, in: Lakatos, I. & Musgrave, A. (eds) *Criticism and the Growth of Knowledge*, Cambridge: Cambridge University Press, pp. 51-58.

Popper, K.R. (1972) Objective Knowledge: An Evolutionary Approach. Oxford: Clarendon Press.

Popper, K.R. (1976) *Unended Quest*, Illinois: Open Court Publishing.

Pradeu, T. (2014) Regenerating theories in developmental biology, in: Minelli, A. & Pradeu, T. (eds) (2014) *Towards a theory of development*, Oxford: Oxford University Press, pp. 15-32.

Prigogine, I. (1977) *Self-organization in Non-equilibrium Systems*, New York: Wiley Interscience.

Prigogine, I. (1987) Exploring complexity, *European Journal of Operational Research*, vol. 30, pp. 97-103.

Prigogine, Ilya (2000) The Future Is Not Given, in Society or Nature, *New Perspectives Quarterly*, vol.17, no. 2, pp. 35-37.

Pigliucci, M., & Müller, G.B. (2010) *Evolution, the extended synthesis*, Cambridge, MA: MIT Press.

Quine, W.O. von (1953) Two Dogmas of Empiricism, in: idem, *From a Logical Point of View*, London/New York: Harper & Row, pp. 20-46.

Raad voor Ruimtelijk- Milieu- en Natuuronderzoek (2005) *Interdisciplinariteit en beleidsrelevantie in onderzoeksprogramma's: een stellingname*, Den Haag: Raad voor Ruimtelijk, Milieu- en Natuuronderzoek (RMNO), June 2005.

Raskin, P. (2008) World lines: a framework for exploring global pathways, *Ecological Economics*, vol. 65, pp. 461-470.

Raskin, P., Banuri, T., Gallopin, G., Gutman, P., Hammond, A., Kates, R., Swart, R. (2002) *Great transition. Umbrüche und Übergänge auf dem Weg zu einer planetarischen Gesellschaft*, Materialien Soziale Ökologie, vol. 20.

Reason, P. & Bradbury, A. (eds) (2001) *Handbook of Action Research: Participative Inquiry and Practice*, London: Sage.

Reed, M. & Harvey, D.L. (1992) The new science and the old: Complexity and realism in the social sciences. *Journal for the Theory of Social Behaviour*, vol. 22, no. 4, pp. 353-380.

Regeer, B. & Bunders, J. (2007) *Kenniscocreatie: samenspel tussen wetenschap & praktijk. Complexe, maatschappelijke vraagstukken transdisciplinair benaderd.* The Hague: COS & RMNO, pp. 11-20.

Repko, A. (2012) *Interdisciplinary Research. Process and Theory,* California: Sage Publications Inc.

Rescher, N. (1998) *Complexity. A Philosophical Overview,* New Brunswick, NJ: Transaction.

Revkin, A. (1992) *Global Warming. Understanding the Forecast,* New York: Abbeville Press.

Richardson, K.A., & Lissack, M.R. (2001) On the status of boundaries, both natural and organizational: a complex systems perspective, *Emergence, A Journal of Complexity Issues in Organizations and Management,* vol. 3, no. 4, pp. 32-49.

Rifkin, J. (2014) *The Zero Marginal Cost Society. The Internet of Things, The Collaborative Commons, and the Eclips of Capitalism,* London: Palgrave & MacMillan.

Rittel, H. & Webber, M.M. (1973) Planning problems are wicked, *Polity,* vol. 4, pp. 155-169.

Robinson, K. (2010) *Changing Education Paradigms,* https://www.youtube.com/watch?v=zDZFcDGpL4U, Uploaded 14 October 2010.

Rockström, J., Steffen, W., Noone, K., Persson, A., Chapin, F.S., Lambin, E.F. et al. (2009) A safe operating space for humanity, *Nature,* No. 461, pp. 472-475. DOI:10.1038/461472a.

Rogers, K., Luton, R., Biggs, H., Biggs, R., Blignaut, S. Gholes, A., Palmer, C. & Tangwe, P. (2013) Fostering Complexity Thinking in Action Research for Change in Social-Ecological Systems, *Ecology & Society,* vol. 18, no. 2:31.

Rorty, R. (1989) *Contingency, Irony and Solidarity,* Cambridge: Cambridge University Press.

Rushdie, S. (1990) *Haroun and the Sea of Stories,* London: Granta Books.

Saam, N. (2017) What is a Computer Simulation? A Review of a Passionate Debate, *Journal of General Philosophy of Science,* vol. 48, pp. 293-309.

Sadava, D., Heller, H.C., Orians, G.H., Purves, W.K. & Hillis, D.M. (2014) *Life. The Science of Biology,* Sunderland: Sinnauer Associates Inc.

Scheffer, M., Carpenter, S., Lenton, T., Bascompte, J., Brock, W., Dakos, V., Koppel, J. van de, Leemput, I. van de, Levin, S., Nes, E. van, Pascual, M. & Vandemeer, J. (2012) Anticipating Critical Transitions, *Science,* vol. 338, 19 October 2012, pp. 344-348.

Schön, D. (1987) *Educating the Reflective Practitioner. Toward a New Design for Teaching and Learning in the Professions,* San Fransicos/London: Jossye-Bass Publishers.

Senge, P. (2005) *The Fifth Discipline. The Art & Practice of the Learning Organization,* New York/London etc: Currency Doubleday.

Shackley, Simon, Wynne, Brian, Waterton, Claire (1996) Imagine complexity: the past, present and future potential of complex thinking, *Futures*, vol. 28, pp. 201-225.

Sloot, P. (2016) Ware interdisciplinariteit vereist een nieuwe denkwereld, een ander paradigma, *Nieuwsbrief Instituut voor Interdisciplinaire Studies*, 14 June 2016

Smajl, A., Pech, S. & Hanoi, J.W. (2012) *The Water, Food & Energy Nexus: Results of a Mekong investigation*, CSIRO Ecosystem Science/Climate Adapatation Flagship, November 2012, retrieved from www.slideshare.net/CPWFMekong/ hanoiintroduction-and sessions-v4)

Smajl, A., Ward, J. & Pluschke, L. (2016) The water–food–energy Nexus – Realising a new paradigm, *Journal of Hydrology*, vol. 533, pp. 533-540.

Smith, E. (2015) Corporate Imaginaries of Biotechnology and Global Governance: Syngenta, Golden Rice, and Corporate Social Responsibility, in: Jasanoff, S. & Kim, S.-H. (eds) (2015) *Dreamscapes of Modernity: Sociotechnical Imaginaries and the Fabrication of Power*, University of Chicago Press, pp. 254-276. See also: chapters 1, 13 and 15.

Snow, C.P. (1959 & 1963) *The Two Cultures*, Cambridge: Cambridge University Press.

Spaargaren, G., Oosterveer, P. & Loeber, A. (2012) *Food Practices in Transition. Changing Food Consumption, Retail and Production in the Age of Reflexive Modernity*, New York/London: Routledge.

Steffen, W., Richardson, K., Rockström, J., Cornell, S.E., Fetzer, I. Bennett, E.M. et al. (2015) Planetary boundaries: Guiding human development on a changing planet, *Science*, vol. 347, no. 6223, pp. 730-735. DOI: 10.1126/science.1259855

Sterling, S. (2004) Higher education, sustainability, and the role of systemic learning, in: Corcoran, P.B. & Wals, A.E.J. (eds), *Higher Education and the Challenge of Sustainability*, Dordrecht: Kluwer Academic Publishers, pp. 49-70.

Sterling, S. (2007) From the Push of Fear, to the Pull of Hope: Learning by design, *Southern African Journal of Environmental Education*, vol. 24, pp. 30-34.

Sterling, S. (2009) Sustainable education, in: Gray, D., Colucci-Gray, L., Camino, E. (eds), *Science, Society and Sustainability: Education and Empowerment for an Uncertain World*. New York: Routledge, pp. 105-118.

Sterling, S. (2011) Transformative Learning and Sustainability: sketching the conceptual ground, *Learning and Teaching in Higher Education*, no. 5, pp. 17-33.

Suárez, M. (2004) An inferential conception of scientific representation, *Philosophy of Science*, vol. 71, no. 5, pp. 767-779.

Suárez, M. (2003) Scientific representation: Against similarity and isomorphism, *International Studies in the Philosophy of Science*, vol.17, pp. 225-244.

Suárez, M. (2010) Scientific representation, *Philosophy Compass*, vol. 5, no. 1, pp. 91-101.

Swart, R., Raskin, P. & Robinson, J. (2004) The problem of the future: Sustainability science and scenario analysis, *Global Environmental Change*, vol. 14, pp. 137-146.

Thompson Klein, J. (2004) Interdisciplinarity and complexity: An evolving relationship, *E:CO Special*, vol. 6, nos. 1-2, pp. 2-10.

Thorèn, H. & Persson, J. (2013) The Philosophy of Interdisciplinarity: Sustainability Science and Problem-Feeding, *Journal of General Philosophical Science*, vol. 44, pp. 337-355.

Törnberg, P. (2014) *Worse than complex* (dissertation), Göteborg: Chalmers University of Technology.

Toulmin, S. (1972) *Human Understanding: The Collective Use and Evolution of Concepts*, Princeton, NJ: Princeton University Press.

Toulmin, S. (1990) *Cosmopolis. The Hidden Agenda of Modernity*, New York: The Free Press.

Toulmin, S. (2001) *Terug naar de rede*, Kampen: Agora.

Tromp, C. (2012) The Diffusion & Implementation of Innovation, *Innovative Studies: International Journal*, vol. 2, no. 1, pp. 18-30.

Tromp, C. (2013) Beyond "Business as Usual". A Truly Rational Approach to Smart Technology Implementation, *The European Financial Review*, June/July 2013, pp. 11-18.

Ulanowicz, R.E.A. (2009) *Third window: natural life beyond Newton and Darwin*, West Conshohocken Pennsylvania, USA: Templeton Foundation Press.

Urry, J. (2005) The Complexity Turn, *Theory, Culture & Society*, vol. 22, no. 5, pp. 1-14.

Urry, J. (2005) The Complexities of the Global, *Theory, Culture & Society*, vol. 22, no. 5, pp. 235-254.

Verschoor, G. (2009) Deploying critical capacity, in: Almekinders, C., Beukema, L., Tromp, C. (eds), *Research in action. Theories and practices for innovation and social change*, Wageningen: Wageningen Academic Publishers, pp. 141-155.

Visscher, M., Bodelier, R. (red.) (2017) *Ecomodernisme: het nieuwe denken over groen en groei*, Amsterdam: Uitgeverij Nieuw Amsterdam.

Visser, W. (2015) *Sustainable Frontiers. Unlocking Change through Business, Leadership and Innovation*, Sheffield: Greenleaf Publishing.

Wallace, M.L., Walker, J.D., Braseby, A.M., & Sweet, M.S. (2014) "Now, what happens during class?" Using team-based learning to optimize the role of expertise within the flipped classroom, *Journal on Excellence in College Teaching*, vol. 25, no. 3-4, pp. 253-273.

Wallace-Wells, D. (2017) The Uninhabitable Earth, *New York Magazine*, 9 July 2017.

Walliman, N. (2005) Critical Realism, uit: idem. *Your Research Project*, London etc.: SAGE, pp. 208-209.

Wals, A. & Jickling, B. (2002) "Sustainability" in higher education: From doublethink and newspeak to critical thinking and meaningful learning, *Higher Education Policy*, vol. 15, pp. 221-232.

Wang, X., Su, Y., Cheung, S., Wong, E. & Kwong, T. (2013) An exploration of Biggs' constructive alignment in course design and its impact on students' learning approaches'. *Assessment & Evaluation in Higher Education*, volume 38, issue 3, pp. 477-491.

Webster, K. (2007) Hidden Sources: Understanding Natural Systems is the Key to an Evolving and Aspirational ESD, *Journal of Education for Sustainable Development*, vol. 1, no.1, pp. 37-43. Retrieved 9 August 2016 from jsd.sagepub.com

Widdershoven, G. (1987) *Handelen en rationaliteit – Een systematisch overzicht van het denken van Wittgenstein, Merleau-Ponty, Gadamer en Habermas*, Meppel/ Amsterdam: Boom.

Wiek, A., Withycombe, L. & Redman, C. (2011) Key competencies in sustainability: A reference framework for academic development, *Sustainability Science*, vol. 6, pp. 203-218.

Wiek, A., Bernstein, M., Laubichler, M. Caniglia, G., Minteer, B. & Lang, D. (2013) A Global Classroom for International Sustainability Education, *Creative Education*, vol. 4, no. 4A, pp. 19-28.

Wiek, A., Bernstein, M., Foley, R., Cohen, M., Forrest, N., Kuzdas, C., Kay, B., & Withycombe Keeler, L. (2015) Operationalising competencies in higher education for sustainable development, in: Barth, M., Michelsen, G., Rieckmann, M., Thomas, I. (eds) (2015) *Handbook of Higher Education for Sustainable Development*, London: Routledge, pp. 241-260.

Williams, A. & Srnicek, N. (2013) Accelerate Manifesto for an Accelerationist Politics, *Critical Legal Thinking*, 14 November 2013. http://criticallegalthinking. com/2013/05/14/accelerate-manifesto-for-an-accelerationist-politics

Wilmsen, C. (2008) Extraction, Empowerment, and Relationships in the Practice of Participatory Research, in: Boog, B., Preece, J., Slagter, M. & Zeelen, J. (eds), *Towards quality improvement of action research: Developing ethics and standards*, Rotterdam: Sense Publishers, pp. 7-28, 135-146.

Winsberg, E. (2010) *Science in the age of computer simulation*, Chicago: University of Chicago Press.

Wijffels, H. (2012) *Formeren is vooruitzien. Duurzame hervorming in 21 stellingen*, Utrecht: Barend Toet/Media Minz & VLB Uitgeefgroep.

Wooldridge, M. & Jennings, N.R. (1995) Intelligent agents: theory and practice, *Knowledge Engineering Review*, vol. 10, pp. 115-152.

World Economic Forum (2016) *The Future of Jobs – Global Challenge Insight Report*, https://www.weforum.org/reports/the-future-of-jobs, retrieved d.d. 29 August 2016.

Yolles, M.I. (1996) Critical Systems Thinking, Paradigms, and the Modelling Space, *Systems Practice*, vol. 9, no. 6, pp. 549-569.

Glossary and Index

Words in italics are discussed separately.

Abduction
A way of non-valid logical reasoning where inferences are based on concomitance, on co-occurances or striking similarities in behavioural patterns between the situation at hand and other, comparable situations. [2.2.1; 2.3.2]

Actor causality
A person's autonomous position in relation to most (if not all) surrounding factors; every subject is an autonomous causal source of their own actions in the sense that the influence of environmental factors on individual actions is ultimately – and therefore totally – determined by the subject (Coenen 1987: 143). [3.3.2; 5.4]

Alternative hypothesis
A hypothesis proposed in addition to the *null hypothesis* which states that a specific effect that was to supposed to result from a given cause will not occur. An alternative hypothesis states that a specific effect described as expectation derived from a theory will indeed occur. If the null hypothesis can be rejected, the alternative hypothesis is a *plausible* explanation. [2.1.2]

Anthropocene
A proposed term for the present geological age in which human intervention has had a significant impact on the Earth's systems (geology, ecosystems, climate, atmosphere). [1.1; 3.3.2; 6.1.1; 6.2.2]

Arbitrary stop
An argument stopped at a random point by identifying a particular method or theory as the touchstone or basic proof (Baynes etal. 1987: 251, 302; Smaling 1987: 171). See also *dogmatic stop* [1.3.1; 4.4.2]

Argumentation
A process of reasoning in which participants try to persuade each other with motives and reasons of the truth, *validity* and authenticity of certain propositions. In this process, participants may be persuaded by formal logic (e.g. *deduction*, or the less pure forms of *induction* or *retroduction*) and informal logical diversions (e.g. *rhetorical* devices). [1.2.2 ff.]

Assumption

A basic principle accepted before making logical deductions or starting research, see also *presuppositions* (Toulmin 1990: 116-117) [1 ff.]

Atomistic approach

An approach in which it is considered possible to divide and study a research object in separate independent sections or factors. [1.2.4; 1.2.5; 2.3.3]

Buffer (capacity)

A stock that is big, relative to its inflows and outflows, and thus has some capacity to stabilise itself (Meadows 1999: 7). [1.1; 5.2.2]

Cellular automata

Simulations that assign a discrete state to each node of a network of elements and assign rule of evolution for each node based on its local environment. Such simulations are especially common in the social sciences, where each node can be thought of as an agent reacting to its local environment (Winsberg, 2010: 5). [2.3.1]

Coherence theory of the truth

The idea that the truth of knowledge depends on the degree to which statements about reality form a collective, consistent, coherent whole and provide a reliable explanatory theory. [3.2.3; 3.2.4; 5.2; 5.2.2]

Common sense

Everyday knowledge; interpretations, categories and expectations taken for granted and considered true by social actors regarding the physical and social causal composition of the world. [2.2.3; 4.1; 5.2.1; 5.4; 6.3.1]

Complex Adaptive System

Networks or collections of interdependent agents linked in non-linear ways that display a *buffer capacity* and resilience due to their ability to adapt and organise (Morin 2008; Midgley, 2000: 39-44; Homer-Dixon, 2011). [1.1; 3.2.1]

Complexity (thinking)

A way of thinking that takes account of the idea of *systems thinking*, thus incorporating subtle interactions and retroactions (feedback loops) between different actors and factors that link phenomena in a wider whole, and of the fact that complex issues are consequently characterised by challenging phenomena such as *non-linearity*, *self-organisation* and *emergence*. [1. ff]

Confirmation holism

The idea that it is impossible to test an independent hypothesis because scientific theories can only be evaluated in relation to the whole interdependent structure that each separate assumption is part of. This problem is also known as the Duhem-Quine thesis (Quine 1953). [2.1.3]

Constituent
A basic part of a whole that is created by the different elements it comprises. [1.2.4; 1.2.5; 1.4.2; 3.2.2]

Construction / Constructivism
A theoretical perspective in which attention is focused on how social actors actively construct social and physical reality in social practice and in their day-to-day activities, which forms the basis for their subjective descriptions and explanations of reality. [2. ff.]

Correspondence theory of the truth
The idea that the truth of knowledge lies in its core on the agreement between elementary statements made by knowing subjects and objective reality. Correspondence rules determine the relation between theoretical language (statements by subjects) and observational language (objective reality). [3.1.1; 3.1.3; 3.2.1; 3.2.4; 5.2; 5.2.1]

Corroboration
The degree of resistance that a hypothesis or theory shows against attempts at refutation. [5.2]

Counter-Enlightenment
In contrast to the universalist **Enlightenment** ideal of reason, thinkers of the Counter-Enlightenment emphasised emotion and imagination, which they saw as an expression of a particularist awareness in a specific time and culture (Leezenberg & De Vries, 2005: 120) [4.4.1]

Critical rationalism
The view that scientific knowledge acquisition should be based on a critical approach, and that testing by falsification encourages scientists to produce ever-better theoretical models coming ever closer to the truth (Popper 1934/1959, 1963). [2.1.2; 2.1.3; 2.2; 2.2.2; 2.2.4; 2.4; 3.2 3.2.3; 3.3.2; 4.1; 4.3.1; 5.2.2; 6.2.4]

Critical realism
The view that while theoretical statements may point to an objective reality that exists independent of us, we can never know this reality directly and thus cannot appeal to objective reality or any **correspondence theory of the truth** to justify our knowledge claim. [3.3.2; 3.3.3]

Deconstruction
Literally: the disintegration of a whole into its parts. **Postmodernists** propose deconstruction as a method of dealing with scientific and philosophical exposés, **discourses** or **narratives**: to show that truth is constructed, the explanation or argument is unravelled, demonstrating the assumptions that implicitly support it. [2.2.2]

Decoupling
The idea that, even as human environmental impacts continue to grow in the aggregate, human well-being can be disconnected from environmental destruction. Decoupling can be driven by both technological and demographic trends and usually results from a combination of the two (Asafu-Adjaye et al. 2015; Visscher & Bodelier 2017). [6.2.2]

Deduction
The reasoning by which a conclusion is drawn from certain **premises** based on valid logical principles. Deduction allows specific expectations or predictions (**hypotheses**) to be drawn from general theories or knowledge claims. [1.2.2 ff.]

Deductive-nomologic model (D-N model)
An explanatory model designed by Hempel and Oppenheim in which the term **nomological** refers to the fact that at the top of the model is a general empirical statement or rule. The **deductive** aspect is that from this general empirical rule, using logic and certain basic conditions (the so-called **premises**), specific statements can be inferred (Hempel 1965: 345; 't Hart 2001: 120; Potter 2000: 232). [2.1.2; 2.1.3]

Discourse (Diskurs)
The stories, narratives, and general way of discussion and communication that form the basis for social interaction in a certain time, culture, or context (Foucault 1976, 1979, 1980; Habermas 1981; Lyotard 1984). [2.2.2; 6.2.3]

Deroutinisation
Distancing from established customs and traditions, from the everyday interactions we take for granted to break through the often unconscious reproduction of social structures (Giddens 1979: 220). [6.2.3]

Design thinking
A form of solution-based thinking in which systems thinking is specifically used to develop creative strategies with the intent of producing constructive future results. [2.3.2; 2.3.3; 3; 3.4.2; 5.1.2; 5.4; 6.3.1]

Determinism
The idea that every event or situation is linked to previous events by universal laws of causality that govern the world. [3.1.2; 3.3.1; 6.2.2]

Dogmatic stop
An argument stopped at a random point by identifying a particular method or theory as the touchstone or basic proof (Baynes et al. 1987: 251, 302; Smaling 1987: 171). See also **arbitrary stop**. [1.3.1; 4.4.2]

Dualism /Dualistic
A dichotomy: two different concepts presented as opposites (e.g. subject – object, action – system). [5.1; 5.4; 6.2.3; 6.2.4]

Duality
The idea that two different concepts are their mutual extension; they are not opposites but two sides of the same phenomenon (e.g. with their actions, social actors continuously produce and reproduce social structures, so actions and structure constantly interplay) (Giddens 1985). [3.3.1; 3.3.2; 5.4; 6.2.3; 6.3.1]

Double hermeneutic problem
The double interpretative challenge social scientists face when they want to study social phenomena: 1) their understanding of the phenomena depends on interpretation of the actions of their subjects, 2) while interpreting these actions, they rely on the same kind of regular interpretive frameworks as the actors themselves are relying on (Giddens 1976: 155) [2.2.3]

Enlightenment
A rationalisation process that began as a movement to demystify the world ('disenchantment'), so that life would be seen and interpreted less through religious or biblical worldviews and more through rational explanations showing how the world works. [4.1; 4.4.1; 6.1.1; 6.2.2]

Emergence
The idea that complex organised systems can reveal certain characteristics as they develop which cannot be explained as a combination of composite parts. Higher order structures in complex systems do not flow directly from structures that are their ontological predecessors. Emergence refers to a process of continual development in which different structures within a *system* interact and influence each other in complex ways. This interaction can result in entirely new characteristics, patterns, regularities or entities (Prigonine 1977). [1.1 ff]

Empirical cycle
A process in scientific research in which theories are formed through *induction* via observation, followed by predictions that can be made through *deduction*, and these are subsequently tested and evaluated. This evaluation may lead to new theory formation, thus leading to yet another research circle. [2.1.1; 2.1.2; 2.2.1; 3.4.1; 5.1]

Empirical realism
The idea that the world – or reality – can be observed and experienced as it really is. [3.1.1; 3.1.3; 3.2.1; 3.3.2]

Empiricism
A scientific philosophical movement in which sensory experiences or empirical observations are viewed as the principal source of knowledge. [1.2.1; 1.2.2; 2.1; 4.1]

Epistemology

Knowledge theory, i.e. theory about how people are able to know reality, how we can gain adequate knowledge and what is the best strategy for justifying our ideas about reality. [1.3.2; 3.2; 3.3.3; 5.3.1; 5.3.2]

Ethnomethodology

An approach that focuses on how so-called 'ordinary people' with ordinary common sense develop and use knowledge. The focus is on the mutual interaction between people; the essential elements of this interaction are described, such as the difference between statement and intent and the importance of a shared definition of a situation. [2.2.2]

Event causality

Causal influences linked to conditions and circumstances of which actors are often oblivious or unaware, although these influence their actions (Giddens 1976: 154-5, 159). [3.3.2; 5.4]

Extrapolation

Making anticipatory statements, based on received data, about something regarding which no information is as yet available. The expansion of a set of numbers beyond the existing series. [2.1.3; 5.2.2; 6.2.2]

Fallibilism

A philosophical position that states that knowledge can be refuted and that nothing is known for certain, even if there is good reason to assume it is true. Acknowledging that it is impossible to claim to know everything, this position engenders a certain humility regarding what can be known (Fay 1996: 208). [3.1.2; 3..2.1; 3.2.4; 3.3.2; 5.3.1]

Falsification

A method to refute theories by confronting them with opposing empirical examples. Falsification is the counterpart of **verification**, a method of confirming theories empirically, which is not waterproof from a logical point of view (Popper 1959, 1963, 1968). [2.1.1 ff.]

Fragmentation

The disappearance of unity in everyday consciousness due to the development of domains such as science, law and economics into specialised fields in today's increasingly complex world (Habermas 1981a vol. II: 483, 522). [4.2.2; 5.1.2]

Half-modern

Characterisation of the condition of today's society by some social theorists who analyse that: 1) the ideals of freedom, equality and fraternity still have not been realised, 2) the core driving force behind social development is still a narrow rather than a broad conception of rationality, and 3) there is still not enough attention for the unintended, negative consequences of the modernisation process (Beck 1986; Beck, Giddens & Lash 1994). [4.4.1; 6.2.1; 6.3.1]

Hermeneutics
Originally the study of interpretations of written texts, especially on literature, religion and law. Later this interpretative method was also used within the domains of philosophy and science, encouraged especially by Gadamer. [2.2.1; 2.3.1; 3.4.1; 3.4.2; 5.1; 5.2.1]

Holistic approach
Counterpart to the *atomistic approach*, based on the idea that the whole is more than the sum of its parts, or that the whole has a different quality and has a decisive impact on its consistent parts ('holos' is 'whole' in Greek). [1.2.5; 2.4; 5.1]

Humanities
The study of how people understand and record our world, how they process and document the human experience in philosophy, literature, religion, art, music, history and language. [1.2.4 ff]

Hypothesis
An explicit expectation or prediction that is formulated to be able to test a theoretical explanatory model. [1.2.3 ff]

Ideographic
An approach that focuses on the supposedly unique and special characteristics of people's actions and products. [2.2.1]

Incommensurable / Incommensurability
The idea that scientific theories or paradigms are theoretically irreconcilable since it is impossible to interpret a specific theory in terms of a neutral meta-language, and thus it can only be interpreted in terms of another theory (Kuhn 1962). The assumption here implies a *relativist* view of the scientific process: where a common standard is lacking and a solid comparison cannot be made for some theories or paradigms, it may not be possible to state which paradigm is better or in which theory the *verisimilitude* is greater, or indeed when there is scientific progress (Feyerabend 1970: 219-20). [5.2.2; 5.2.3]

Induction / Inductivism
Method of reasoning in which a *universal* statement is made based on observed regularity in a certain number of cases so that this regularity is deemed to apply in every case (Koningsveld 1987: 202). [1.2.2 ff.]

Induction problem
The problem that the method of reasoning in which a more general theoretical statements is inferred from a collection of particular singular observations is a quasi-logical form that does not provide solid legitimacy for the inference ('t Hart 2001: 113). [1.2.2]

Infinite regress

The danger that while searching for an ultimate foundation it may be necessary to go even deeper to the foundation under the foundation and so forth. The justification shifts ever further to so that an absolute foundation is never achieved (Baynes et al. 1987: 251, 302; Smaling 1987: 171). [1.3.1]

Interpretivism / Interpretivist approach

A movement in which understanding human behaviour (**Verstehen**) is the central point of attention. Interpretivists try to understand the world and the way people act in the world from the individual's perspective. It is about empathically comprehending human behaviour, which is internally driven by certain motives and intentions. The idea is that such understanding is needed to come up with suitable explanations of human actions, especially since people are supposed to be more than the plaything of external forces. [2.2 ff]

Institutionalisation

The process in which frameworks of meaning, norms and values patterns, power relations and codes of conduct are anchored in social institutions. [3.3.1; 6.2.3; 6.3.1]

Instrumentalism / Instrumentalist

Scientific position in which scientific theories are conceived as instruments for making predictions (rather than as providing explanations about the structure of reality, cf. **realism**). Or: vision in which scientific theories and the resulting knowledge is seen as equivalent of the instruments used to test acquired knowledge (Lakatos 1978: 106; Popper 1963: 104, 116). [3.1.3; 3.3.3; 5.2.1]

Instrumental rationality

A definition of rationality in which only 'pure' knowledge counts, i.e. knowledge that is based on our cognitive abilities and logically derived theories. [4.2.1; 4.2.2; 4.4.1; 4.4.2; 4.5; 6.2.4; 6.3.2]

Interdisciplinarity

The combination and integration of different knowledge domains and perspectives. [1.4.1 ff]

Isomorphism

A form of representation with rather strict demands such as that there be a one-to-one function that maps all the elements in the model's structure onto the elements in the real-life structure and vice versa, while preserving the relations defined in each structure (Suárez 2003, 2004: 227-228, 2010: 95; see also Knuuttila 2005: 1261). [3.1.1; 5.2.1]

Landscape

External environmental factors that may aid or obstruct the emergence of a new socio-technical **regime**, the realisation of **transition** in a given **system** to instigate a **paradigm** shift (Grin et al. 2010: 23). [6.2.3; 6.3.1]

Leverage point

A place in a system where there is an opportunity to change its functioning in order to transition the system as a whole to a more sustainable state (Meadows 1999, 2008; Kennedy et al 2018: 8). [6.2.3]

Life politics

The phenomenon of modern individuals engaged in a self-realisation process having to take ever more decisions regarding their own life. In a society in which they are no longer bound to traditional role and expectation patterns, an identity is no longer a given but has to be continually realised and affirmed. This requires a reflexive sense of awareness of a person's own life story (Giddens 1991: 1-5, 9, 52-53). [4.2.2; 6.3.1]

Logical circularity

The risk when justifying an argument of having to base this on a premise or explanation that itself has to be justified. In effect, this may mean continuing in a circle and never being able to find the ultimate absolute foundation (Baynes et al. 1987: 251, 302; Smaling 1987: 171). [1.3.1]

Metaphysics

Originally literally: what lies behind material reality or rises above it. [3.3.3; 4.3.1]

Method

The rules and techniques used to collect date relevant to a theory, e.g. through observations or interviews (Bij de Weg 1996: 63; Harding 1987: 2). [1.2.2 ff.]

Methodology

A theory or analysis about how research should be done; a system of logic or philosophical principles to structure theory-relevant data (Harding 1987: 2; Bij de Weg 1996: 54). [1.3.2 ff]

Modernisation / Modernity

Social development processes that have taken place over the last couple of centuries and that are characterised by: 1) increasing differentiation in a range of societal domains, leading to division of labour, capital formation and increasing productivity, as well as the institution of centralised power and of nation-states, the advent of the right to political participation, urban lifestyles and formal schooling;
2) secularisation: the loss at a cultural level of traditional norms and lifestyles once taken for granted (Beck 1986: 206; Habermas 1984: 10; Korthals & Kunneman 1992: 125). [4.1; ff.]

Modus ponens

A logical argument also referred to as *syllogism* or universal affirmative; a logically valid reasoning in which a major *premise* and one or more minor *premises* form the basis for a conclusion. [1.2.2]

Monism

Declaring a specific method (usually the standard scientific method) as the only valid explanatory model in advance, independently of the nature of the situation, problem or event. [2.4]

Multidisciplinarity

The combination of different knowledge domains and perspectives. [1.4.1; 5.3; 5.4; 6.3.2]

Münchhausen trilemma

A triple choice alternative when attempting to find an absolutely certain foundation to support acquired knowledge whereby none of the options is acceptable. This involves the risks of *logical circularity*, *infinite regress* and the recourse to absolute certainty whereby the justification randomly stops (*arbitrary* or *dogmatic stop*). Albert calls the trilemma the Münchhausen trilemma because the solution is similar to Baron von Münchhausen's attempt to pull himself out of the swamp by his own hair. It is also known as the bootstrap approach (Baynes et al. 1987: 251, 302; cf. Smaling 1987: 171). [1.3.1]

Narrative

A tale, interpretive, with the form or structure of a (life) story. Social actors interpret the world as stories in which they take part or act. Narrative structure and storytelling are two sides of what constitutes reality, referring to the primary foundation for meaning. As a cognitive process, the creation of narratives organises experience into temporal episodes. We also combine theories about reality in a story that is as coherent and consistent as possible (cf. the *coherence theory of truth*). [1.4.2; 2.2.2; 3.2.3; 4.2.2; 4.4.2; 6.2.1; 6.3.1]

Naturalism

The view that the natural scientific method can be used as model in the social sciences too, problem-free, since there is no reason to suppose that the social sciences are confronted with fundamentally different research objects. (Giddens 1985: 27-28). [1.2.4; 1.3.3; 5.2.2]

Niche

The recesses of a *system* in which technical and social innovations can come into existence and further develop, bottom-up (Grin et al. 2010: 22-23; Spaargaren et al. 2012: 4-6). [6.2.3; 6.3.1]

Nomologic / Nomothetic

General rules relating to causal connections. [1.2.4; 1.2.5; 2.1.2; 2.1.3; 2.2.1]

Null hypothesis
The hypothesis that states that no effects will be found as a result of a particular cause attributable to a certain function. Scientific research is structured in such a way that the null hypothesis can hopefully be rejected, to lend more **plausibility** to the **alternative hypothesis**. [2.1.2]

Objectivism / Objectivist
Viewing phenomena as objects in the belief that a permanent, non-historical frame of reference exists that governs what **rationality**, knowledge, truth, correctness, etc. are (Smaling 1987: 175). [3.1.3 ff.]

Ontology
Theory about reality; theory about the principles that constitute and structure reality. [1.3.2; 1.3.3; 3.2; 6.2.4]

Operationalisation
The conversion of theoretical terms from expectations into empirically verifiable statements, or the conversion of a concept into a measurable construct or instrument to enable theories to be tested. [1.2.5; 2.4]

Orthodox consensus
A widely shared common view, certainly until the 1960s yet still dominant today, regarding the idea that scientific knowledge acquisition can best be modelled after the natural scientific approach to research, and that objectivity and generalisability should be regarded as generally valid and useful criteria. [1.2.4; 1.2.5; 2.2]

Paradigm
A system of assumptions that is taken for granted by a research school or movement and which forms the framework within which solutions to problems are sought (Kuhn 1962, 1970). [1.3.2 ff]

Paradox
A claim that contains an inherent contradiction, i.e. which conflicts internally. [2.1.3; 4.2.2; 4.3.1; 6.1.2]

Perspectivism
The conviction that our view of reality is dictated by our frames of reference and conceptual frameworks. A conceptual framework is a complex of coherent, hierarchically ordered basic assumptions and concepts. These direct our vision and incorporate certain assumptions that may introduce a certain bias regarding what we expect to find in reality. [3.1.2; 3.2.4; 3.3.2; 5.3.1]

Plausible
Probable, probability [2.1.1; 2.1.3; 5.1.3; 6.2.2].

Pluralism

Rejection of **monism**, i.e. declaring a specific method (usually the standard scientific method) as the only valid explanatory model in advance, independently of the nature of the situation, problem or event. [2.4; 5.4]

Postmodern / Postmodernism

A philosophical position in which a shift is proposed beyond modern thinking towards a way of thinking in which Western culture fundamentally changes its perception of itself and of cultural forms such as science and art, knowledge and power. The modernist tendency of wanting to control everything in the world is criticised, and with that the focus on the position of the subject (as source of control), the search for unity (theoretical form of control) and hierarchy, as well as the apparently natural conformity to norms or essentials. By contrast, postmodernists promote **deconstruction**, heterogeneity and pluriformity (Allen 1989: 37; Van Peperstraten 1991: 58). [2.2.2; 4.4.1; 6.2.1]

Postulate

A basic assumption or **constituent** principle that establishes a certain practice or approach to reality from which it is impossible to deviate without abandoning the practice or the approach to reality (Coolen 1992: 42). [1.2.4; 1.2.5]

Premises

Basic assumptions in logical reasoning stating something that is taken to be true and on the basis of which it is possible to reach a conclusion. [1.2.2; 1.2.5; 1.3]

Presupposition

A basic principle accepted before making logical deductions or starting research, see also **assumptions** (Toulmin 1990: 116-117). [2.2; 4.3.1; 5.1.3]

Rationalism

A philosophical movement that emerged in the seventeenth century founded on the idea that reason is the only or the principal source of knowledge. Rationalists argue that reality is inherently reasonable and logical. So it should be possible to 'read' reality just by using our cognitive capabilities, purely through thought, without any intervention or mediation of other means. [1.1.1; 1.2.2; 2.1; 4.4.1]

Rational / Rationality

Knowledge that is acquired by relying on our capacities to reason. In a narrow conception of rationality, knowledge is connected to the senses and reasoning (cognition) and to the need for clarity (i.e. formal logic), accuracy and strictness. So it is hoped to obtain generally valid, timeless and context-free laws. Defined in a wider sense, the concept is held to imply informal reasoning and instructive practical experience too; knowing is not limited to 'facts' but also incorporates norms and values and needs. It is hoped that this way, knowledge can be acquired which may be tied to a time and context yet may still have a wider application in similar circumstances. [1.2.1 ff.]

Reductionism

An approach in which a research object can in theory be dismantled and its operational parts or factors studied separately. [1.2.4; 1.2.5; 2.3.3; 6.2.4]

Reflexive / Reflexivity

The receptivity to (most aspects of) thought and action and associated principles that may be reviewed in light of new knowledge or information and where necessary adjusted (Giddens 1991: 20). [4.2.2; 4.3.2; 4.3.3; 5.3.2; 6.1; 6.1.3; 6.3; 6.3.1]

Reflexive design

A programme to bring *reflexive modernisation* into concrete practice; a way in which to do research and attempt to encourage progress inspired by a particular vision and building upon lessons drawn from the past (Bos & Grin 2008: 482). [6.3.1]

Reflexive modernisation

Process in which the currently held foundations of the social development process are themselves critically discussed and examined for inherent contradictions, thus providing a clear view on their significance for today's society and enabling us to make adjustments where needed (Beck 1986: 253-254). [6.3.1]

Regime

The layer at which systemic change can take place. Once *niche* developments have directed day-to-day practice to other paths (*deroutinisation*) and enough stable new patterns have formed that are embedded and *institutionalised*, this can lead to a new socio-technological *system* or regime (Grin et al. 2010; Spaargaren et al. 2012: 4-6). [6.2.3; 6.3.1]

Regimes of justification

Value-charged frameworks in which debaters raise arguments to support a particular vision or a position regarding a complex issue in a public debate. They try to generalise the value they attach to a particular aspect of the issue (or its solution). They do this by appealing to others to support their generalised argument for the common good. [5.1.3; 5.3; 5.3.3; 5.4; 6.2.4; 6.3.2]

Reification / Reified

Presenting a state of affairs in reality or a model of reality in such a way as to make it seem as though it can be taken for granted, for instance by pretending it represents the 'given' or 'natural' order, and by hiding the fact that it actually implies a social construct. The state of affairs or model is then made into a 'thing' that seems unchangeable (Giddens 1985: 195). [2.3.3; 3.3.1; 5.4]

Relativism

The idea that the way reality is perceived, or the validity of knowledge and statements or the correctness of values and norms is viewed, is wholly or partially dependent on the viewpoints of an individual, a group, community, society, culture, historical period, ecological circumstance, linguistic or conceptual *system*, *paradigm* or worldview. (Smaling 1987: 174-175). [3.2.3; 3.3.2]

Rhetoric

The art of persuasion, the skill of eloquence, the ability to reason clearly and to convince by following the rules of formal logic as well as the rules of everyday, informal logic. In rhetoric, the aim is to ensure that certain propositions are accepted by pointing out their intrinsic positive qualities in a given context, rather than presenting substantive reasons or foundation, as is the case with more formally based arguments (Rescher 2001: 77). [4.4.1; 4.4.2]

Retroduction

A way of non-valid logical reasoning where inferences are based on concomitance, on co-occurrences or striking similarities in behavioural patterns between the situation at hand and other, comparable situations. (See also *abduction*.) [2.2.1; 2.3.2]

Scienticism

The equation of true knowledge with science. [4.3.1]

Self-organisation

The power of living systems and social systems to change themselves utterly by creating whole new structures and behaviours (cf. evolution in biology and technical advance or social revolution in human society) (Meadows 1999: 14). [1.1; 1.2.5; 2.3.3; 5.1.1]

Syllogism

A logically valid reasoning, also called *modus ponens*, in which a major *premise* and one or more minor *premises* form the basis for a conclusion. [1.2.2]

System / Systemic

An entity of interrelated and interdependent parts that is coherently organised and interconnected in a pattern or structure that produces a characteristic set of behaviours, often classified as its 'function' or 'purpose' (Meadows et al. 2008). [1.1 ff]

Systems thinking

The interdisciplinary study of *systems*, including the subtle interactions and retroactions (feedback loops) between different actors and factors that link phenomena in a wider whole. [1.3.3; 2; 2.3; 2.3.2; 6.2.3]

Target system

The real-life object, phenomenon, process or system in which scientists are interested when they build a model to represent it in some form or other (Knuuttila 2011: 264). [2.3.1; 2.3.2; 3.1.1; 3.1.3; 3.2.3; 5.1.1; 5.2.2; 5.3.1]

Transdisciplinarity

The combination and integration of different knowledge domains and perspectives, including the perspectives of stakeholders outside academic life. [1.4.1 ff]

Transition

A process of structural change brought about by new ideas, technologies, products, frameworks, *discourses*, infrastructures, practices and patterns from which a new *system* can emerge (Grin et al. 2010; Spaargaren et al. 2012: 4-6). [1.3.2; 4.2.2; 6.1.2; 6.2.3; 6.2.4; 6.3.1; 6.3.2]

Unified science

The belief that it is possible to strive for unity of the natural sciences and social sciences, expressed in the desire for a uniform scientific language and a uniform scientific method. [1.2.4; 1.3.3]

Validity / Validation

The extent to which concepts, theories and methods, and collected data, results and research conclusions can be considered adequate and have explanatory value with respect to the phenomena that scientists set out to explain. [1.2.2 ff]

Valorisation

To give knowledge a social value by making it useful in society. [4.3.3; 6.1.2].

Verification

The attempt to confirm predictions (*hypotheses*) through confrontation with empirical observations, making theories more plausible. The emphasis here is on the positive function of argument. [1.2.2; 2.1.1; 2.1.2; 2.3.1; 2.3.2; 3.1.1]

Verisimilitude

The proximity to the truth, whereby relative proximity is based on the truth content minus the falsehood content of (successive) scientific theories plus the degree to which the theory in question is substantiated (which depends on the number of falsification attempts endured) (Popper 1963, 1972; Lakatos 1978: 156). [3.1.1; 5.2]

Verstehen

The concept of understanding that is thought to lie at the heart of the way actors deal with social reality, as it is not just a special technique or method with which to approach social reality but the existential precondition for social existence as such (Gadamer 1960; Giddens 1976: 19, 151; Habermas 1981 vol. I: 158). [2.2.1]

Colophon

About the author

Coyan Tromp teaches Philosophy of Science, Futures Thinking, Interdisciplinary Methodology and 21st Century Skills at the Institute for Interdisciplinary Studies of the University of Amsterdam. She is also curriculum developer, and co-designer of the Interdisciplinary Bachelor's Programme Future Planet Studies and the Minor Science for Sustainability.

About the University of Amsterdam

The University of Amsterdam (UvA) is one of the largest comprehensive universities in Europe, with some 30,000 students, 5,000 staff, and a budget of more than 600 million euros. The University provides academic training in all areas of science and scholarship and welcomes students and staff from all backgrounds, cultures and faiths who wish to devote their talents to the development and transfer of academic knowledge as a rich cultural resource and foundation for sustainable progress.

About the Institute for Interdisciplinary Studies

The Institute for Interdisciplinary Studies (IIS) is the University of Amsterdam's knowledge centre for interdisciplinary learning and teaching. It develops new courses in collaboration with the faculties.

The IIS has more than fifteen years of experience in interdisciplinary education and continuously develops substantive education innovations with an interdisciplinary character. The Institute identifies new themes and issues linked to current developments in academia and society.

Over 3,000 students study at the IIS. The IIS offers a number of interdisciplinary study programmes along with a wide range of electives (minors, honours modules and various public events) for students from any faculty, for staff and for members of the public. All its activities are interdisciplinary in nature and are designed in collaboration with one or more faculties.

About the series

Interdisciplinary education and research is becoming increasingly popular in and outside academia. Yet there is still a demand for a theoretical and practical framework that describes what interdisciplinarity entails and how it can be realised in practice.

The *Perspectives on Interdisciplinarity* series is designed to address these needs and enable universities and curriculum leaders to shape interdisciplinary learning, teaching and research. The books in this series provide students, teachers and curriculum developers with insights into the broad field of interdisciplinary studies, offering practical tools for addressing the challenges that arise when taking an interdisciplinary approach.

The authors and editors who contributed to the publications are all engaged both conceptually and practically in interdisciplinary education and research.

The series welcomes monographs and edited volumes in English and Dutch by both established and early-career researchers, teachers or curriculum developers on topics such as student textbooks for interdisciplinary courses, educational approaches to enhance interdisciplinary understanding, methods for interdisciplinary research, and interdisciplinary theory and methodology.

Previous titles

- Joris J.W. Buis, Ger Post & Vincent R. Visser (2016) *Academic Skills for Interdisciplinary Studies*. Volume 1 of the Series Perspectives on Interdisciplinarity, Amsterdam University Press B.V., Amsterdam.
- Steph Menken & Machiel Keestra (eds) (2016) *An Introduction to Interdisciplinary Research – Theory and Practice*. Volume 2 of the Series Perspectives on Interdisciplinarity, Amsterdam University Press B.V., Amsterdam.
- Linda de Greef, Ger Post, Christianne Vink & Lucy Wenting (2017) *Designing Interdisciplinary Education – A Practical Handbook for University Teachers*. Volume 3 of the Series Perspectives on Interdisciplinarity, Amsterdam University Press B.V., Amsterdam.
- Hannah Edelbroek, Myrte Mijnders & Ger Post (2018) *Interdisciplinary Learning Activities*. Volume 4 of the Series Perspectives on Interdisciplinarity, Amsterdam University Press B.V., Amsterdam.

Contact

Institute for Interdisciplinary Studies
Science Park 904
1098 XH Amsterdam
Tel. +31 20 525 51 90
www.iis.uva.nl
Onderwijslab-iis@uva.nl